Safety Management

Near Miss Identification, Recognition, and Investigation

Safety Management

Near Miss Identification, Recognition, and Investigation

Ron C. McKinnon, CSP

CRC Press
Taylor & Francis Group
Boca Raton London New York

CRC Press is an imprint of the
Taylor & Francis Group, an **informa** business

CRC Press
Taylor & Francis Group
6000 Broken Sound Parkway NW, Suite 300
Boca Raton, FL 33487-2742

© 2012 by Taylor & Francis Group, LLC
CRC Press is an imprint of Taylor & Francis Group, an Informa business

Printed in the United States of America on acid-free paper
Version Date: 2011908

International Standard Book Number: 978-1-4398-7946-7 (Hardback)

Library of Congress Cataloging-in-Publication Data

McKinnon, Ron C.
    Safety management : near miss identification, recognition, and investigation / Ronald C. Mckinnon.
       p. cm.
    Includes bibliographical references and index.
    ISBN 978-1-4398-7946-7 (hardback)
    1. Industrial safety--Management. 2. Industrial accidents. 3. Accident investigation. I. Title.

T55.M387 2012
658.3'82--dc23
                                                           2011035587

Visit the Taylor & Francis Web site at
http://www.taylorandfrancis.com

and the CRC Press Web site at
http://www.crcpress.com

Contents

Preface

INTRODUCTION

Near miss incidents, close calls, or close shaves have often been referred to as "safety in the shadows," as this is where the heart of the accident problem lies. Near miss incidents offer management an opportunity to rectify a system breakdown before it happens. They are inexpensive learning opportunities. Because there are no losses as a result of an undesired event does not necessarily mean that the event is insignificant. Many of these seemingly unimportant events have high potential for injury and other losses. If recognized, reported, and rectified, near miss incident root causes will be eliminated leading to a radical reduction in injury-causing accidents.

MISUNDERSTOOD

For many organizations, the term *near miss* is not only misunderstood, but it is underrated with regards to the potential for a near miss incident to become a profit-draining accident and possible injury at a workplace. The term *near miss incident* also can be defined as a narrowly avoided mishap. What that means in the manufacturing, construction, or mining industry is that a person narrowly avoids an injury due to an unforeseen mishap or when there is an undesired event, which, by a stroke of luck, narrowly avoids damaging a piece of equipment, property, or material. These are missed safety signals.

Reporting and rectifying the causes of near miss incidents has many benefits. Studies of adverse events, such as accidents, indicate that near misses occur more frequently than accidents and are often precursors to accidents. In many cases, the same near miss incident has occurred numerous times prior to the actual accident.

ACCIDENT ROOT CAUSE INDICATORS

Research of thousands of undesired (accidental) events has shown that the outcome of the event cannot be predicted and that, under slightly different circumstances, the consequences could have been better or worse if it were not for some factor of luck or good fortune.

The principle of *multiple causes* indicates that accidents are usually the result of a multitude of causes and there are usually many immediate causes and numerous root causes behind every event.

These loss-producing events are termed *accidents*. Some refer to them as *incidents,* but, for clarity, they will be referred to as *accidents* in this publication. No-loss events with potential for loss will be termed *near miss incidents*.

The high risk acts of a worker or a high risk work environment riddled with hazards, or a combination of both, are the immediate causes or the closest causes of an accident, which results in accidental losses, such as death, injury, property damage, fire, or business interruption. High risk acts and/or conditions are the most obvious accident causes, or the causes that lead to the contact with a source of energy that causes the subsequent loss.

Root, or basic, causes are the deep hidden person and job factors that give rise to the immediate causes in the form of high risk acts and/or conditions. If they are not identified and rectified, the accident problem will not be eliminated. Fixing the immediate causes rectifies the symptom, but not the root or basic cause.

Risk assessment of all near miss incidents will determine which near miss incidents warrant a full investigation to track and eliminate the source of the problem at the root.

A Proactive Approach

S. L. Smith writing in *Occupational Hazards* (1994) says:

> Near miss incidents challenge the tradition of using an accident to initiate a thorough review of safety conditions, practices, and training. Tracking near miss incidents offers organizations a better opportunity to focus their preventative efforts (p. 34).

If based on near miss incident information, these efforts will be proactive rather than reactive. As another safety professional put it:

> Letting a near miss incident go unreported provides an opportunity for a serious accident to occur. Correcting these actions or conditions will enhance the safety within your organization and provide a better working environment for everyone involved. Don't let yourself or co-workers become statistics—report near miss incidents to your supervisor. Prevent an accident that's about to happen!

H. W. Heinrich

More than 80 years ago, H. W. Heinrich suggested that one should focus on the accident rather than the injury. He was the first to propose a ratio existed between injuries and accidents that produced no injuries.

> Accidents and not injuries should be the point of attack. Analysis proves that for every mishap resulting in an injury there are many other similar accidents that cause no injuries whatsoever (p. 24).

Explaining his ratio, the first ever published, he said:

> From data now available concerning the frequency of potential-injury accidents, it is estimated that, in a unit group of 330 accidents of the same kind and involving the same person, 300 result in no injuries, 29 in minor injuries, and 1 in major or lost-time injury (p. 24).

HEINRICH'S THIRD AXIOM

In 1931, H. W. Heinrich drew up a list of 10 axioms based on his safety research, which was published in *Industrial Accident Prevention,* 3rd ed. (McGraw-Hill, 1950). Axiom 3 has great significance for the concept of near miss incidents and he was the first person to derive the following conclusion:

> The person who suffers a disabling injury caused by an unsafe act, in the average case has had over 300 narrow escapes from serious injury as a result of committing the very same unsafe act. Likewise, persons are exposed to mechanical hazards hundreds of times before they suffer injury (p. 10).

His fourth axiom was the first recorded theory that fortune or luck may play a part in determining not only the outcome of an undesired event, but also the severity of consequent injury.

> The severity of an injury is largely fortuitous—the occurrence of the accident that results in injury is largely preventable (p. 10).

Despite these major findings, near miss incidents have mostly been overlooked in industry despite history of major-loss events confirming the theory that there are many near misses or warnings before the occurrence of major accidental losses. Near miss incidents are truly the foundation of major injuries, the building blocks of accidents, and warning signs that loss is imminent.

WEB DOWNLOADS

Additional material is available from the CRC Web site: www.crcpress.com at http://www.crcpress.com/product/isbn/9781439879467

Under the menu Electronic Products (located on the left side of the screen), click on Downloads & Updates. A list of books in alphabetical order with Web downloads will appear. Locate this book by a search, or scroll down to it. After clicking on the book title, a brief summary of the book will appear. Go to the bottom of this screen and click on the hyperlinked "Download" that is in a zip file.

Or you can go directly to the Web download site, which is www.crcpress.com/e_products/downloads/default.asp

Acknowledgments

With so much information and work that has culminated in this book, numerous people need to be thanked. I would like to thank the industrialists and miners that I have dealt with over the past 38 years. I learned industrial safety from them and a great deal about near miss incidents from their near miss investigation systems and from their experiences in dealing with the aspect of "safety in the shadows."

The safety pioneers that I have quoted in this book need to be thanked for their diligent research into one of safety's hidden secrets and for exposing what could be a key to injury reduction at the workplace—near miss recognition, reporting, and rectification. They were the propounders of the important theory that near misses are the foundation of a major injury, and, in modern terms, precursors to major accidental events.

To my many mentors, I thank you for sharing your safety knowledge and for the support and encouragement that you offered me. To my associates, colleagues, and "safety *boytjies*" who I have worked with in many countries, it was a privilege to have met and worked with you. You taught me a great deal.

Thanks to Lisa Nevitt, projects manager, safety and training, Phoenix Water Services Department, City of Phoenix, for the case study and examples of their near miss incident reporting card. Thanks to Chuck Gessner, former Magma/BHP Copper director of safety and loss control, for the Magma Copper case study.

For making this publication possible, I thank my wife, Maureen McKinnon, who spent numerous weeks typing and editing this manuscript. This support warrants my deep gratitude.

The contents of this document are dedicated to the thousands of men and women in industry and mines who have died as a result of occupational injuries and diseases, and to the millions who have been and are injured every year in industries and mines around the world. *Note*: If warnings in the form of near miss incidents had been heeded, I am sure a large number of these accidents could have been prevented.

Every effort has been made to trace rights holders of quoted passages and researched material, but if any have been inadvertently overlooked, the publishers would be pleased to make the necessary arrangements at the first opportunity.

About the Author

Ron C. McKinnon, CSP, is an internationally experienced and acknowledged safety professional, author, motivator, and presenter. He has been extensively involved in safety research concerning the cause, effect, and control of accidental loss, near miss reporting, accident investigation, and safety promotion.

McKinnon received a national diploma in Technical Teaching from the Pretoria College for Advanced Technical Education; a diploma in Safety Management from Technikon SA, South Africa; and a management development diploma (MDP) from the University of South Africa in Pretoria. He also has a master's degree in Safety and Health Engineering from Columbia Southern University in Alabama.

From 1973 to 1994, McKinnon was affiliated with South Africa's National Occupational Safety Association (NOSA) in various capacities, including safety and health training and motivation. He is experienced in implementation of safety programs, safety culture change interventions, auditing safety systems, and production of training films and videos. During his tenure with NOSA, he implemented safety systems and provided training in seven different countries.

From 1995 to 1999, McKinnon was safety consultant and safety advisor to Magma Copper and BHP Copper North America, respectively. At BHP Copper, he was a catalyst in the safety revolution in the copper industry that resulted in an 82-percent reduction in the injury rate and an 80-percent reduction in the severity rate.

In 2001, he spent two years in Zambia introducing the world's best safety practices to the copper mining industry. From there, he accepted a two-year contract in the Kingdom of Bahrain, Arabian Gulf, where he successfully implemented a safety culture change for the country's second largest employer.

After spending two years in Hawaii at the Gemini Observatory, he retired back to South Africa. He now is a consultant to organizations in the United States and is often a keynote speaker at international safety conferences.

McKinnon is the author of *Cause, Effect and Control of Accidental Loss,* published by CRC Press in 2000, and *Changing Safety's Paradigms,* published in 2007 by Government Institutes, USA. He also wrote the book, *Safety and Health at Work: An Introduction,* currently being reviewed for publication.

McKinnon is a professional member of the ASSE (American Society of Safety Engineers), Tucson Chapter past president, and an honorary member of the Institute of Safety Management. He is currently a safety consultant, safety culture change agent, motivator, and trainer.

1 Introduction

Definitions: The definitions of the terms used throughout this book will be repeated in a number of chapters. The reason for this seeming duplication is to clearly explain the concepts so that a clear understanding is given as to what an accident, near miss incident, or other concept is and how it is defined.

CLEARING THE CONFUSION

Experience gained in many organizations internationally has shown that confusion exists within organizations, as well as within the safety and health profession, as to what a near miss incident is and how to identify it in relation to an accident, incident, and unsafe (high risk) behaviors and conditions. This uncertainty has led to near miss incidents being incorrectly labeled and, consequently, almost forgotten.

Some also teach that all near miss incidents must be investigated—an almost impossible and impracticable task. If there is confusion within the minds of safety professionals, that confusion is passed on to employees and management and the end result is that near misses are not recognized, reported, or acted upon. This confusion is possibly the reason for near miss incident reporting systems not existing, or the failed attempts at near miss incident reporting in organizations.

Once understanding is reached as to what exactly a near miss incident is, near miss recognition is much easier. The approach taken in this publication is to keep the concepts simple so that all can understand the difference between the various concepts.

MINOR INJURY IS NOT A NEAR MISS INCIDENT

One event that is often referred to as a near miss incident is an accident that results in minor injury, which could have been a lot worse. This is not a near miss incident; if there is an injury, it is an accident. An accident is the event and the injury is a consequence.

As an example, an operator was splashed with acid from a degreasing process and, because he was wearing the correct personal protective clothing, only received minor acid burns to one arm. Under slightly different circumstances, the injuries could have been more severe if, for instance, his face shield had been out of place or he was not wearing gloves, etc.

The fact that the injury was minor in relation to the potential for serious injury does not rate this event as a near miss incident. It was an accident that resulted in minor injury (loss) and should be termed as such. The fact that there was high potential for serious injury that didn't occur does not qualify this event as a near miss incident. In some instances, an accident scenario could involve injuries, damage, and near miss incidents all in one event.

NEAR MISS INCIDENTS

Near miss incidents are near miss events that come close to causing some form of loss, as there was an actual flow of, or exchange of, energy below the threshold level. In some instances, the flow of energy may have dissipated without making any contact, thus causing no loss. In most cases, the energy does not contact anything, thus causing no harm. In some cases, the exchange of energy was insufficient to cause loss or injury, but the fact that there was an exchange of energy is reason enough to heed the warning. Remember, it's not what happened—but what could have happened. Near miss incidents are accidents *waiting in the shadows.*

DEFINING A NEAR MISS INCIDENT

Near miss incidents are also known as: near miss, or incident, close shaves, or warnings.

Other familiar terms for these events are: *close calls*, or, in the case of moving objects, *near collisions*. Near miss incidents also sometimes have been termed *near hits* by some writers.

A near miss incident is

- An undesired event that, under slightly different circumstances, could have resulted in harm to people, or property damage, or business disruption, or a combination.
- An accident with no injury or loss.
- An event that narrowly missed causing injury or damage.
- An incident where, given a slight shift in time or distance, injury, ill health, or damage easily could have occurred, but didn't this time around.

Merriam–Webster defines a near miss incident as: "A result that is nearly, but not quite, successful." What does this mean to industry? It simply means that a serious accident (loss) *almost* occurred. Someone trips over a pallet, but doesn't fall. Two forklifts *almost* collide at a corner. A tool is dropped, but toes are missed … this time.

Wikipedia, the free online encyclopedia, defines a near miss incident as: "An unplanned event that did not result in injury, illness, or damage, but had the potential to do so." Only a fortunate break in the chain of events prevented an injury, fatality, or damage. Although human error is commonly an initiating event, a faulty process or system invariably permits or compounds the harm and is the focus for improvement.

From here on this book will refer to a near miss incident as: An undesired event, which, under slightly different circumstances, could have resulted in harm to people, or property damage, or business disruption, or a combination of the three. (No substantial loss is experienced.)

DEFINING AN ACCIDENT

There is confusion in the safety and health field concerning the words *accident, incident*, and *near miss incident*. Many years ago, the term *incident* was used to describe *near misses,* but since the modern approach is to term *accidents* (loss producing undesired events) as *incidents,* confusion between *accident* and *incident* still exists.

The American Society of Safety Engineers (ASSE) defines "a near miss accident" as an incident and further defines it as "an undesired event that, under slightly different circumstances, could have resulted in personal harm or property damage; any undesired loss of resources." This definition seems to be a combination of the definitions of an accident *and* a near miss incident and is confusing.

Often the word *accident* is replaced by the term *incident,* which leads to confusion; therefore, this publication will refer to near misses/close calls/near hits as *near miss incidents* to clearly and definitely remove any confusion concerning terminology.

Here are some definitions of an accident to indicate that there is general consensus that an event termed an *accident* results in some form of loss, either to an individual, property, organization, or all of these. These definitions will explain the terminology used throughout this publication.

- An accident is an undesired event often caused by unsafe acts or unsafe conditions and results in physical harm to persons, damage to property, or business interruption.
- An accident is an unplanned, uncontrolled event caused by unsafe acts or unsafe conditions and that results in harm to people or damage to property and equipment.
- An accident is the culmination of a series of activities, conditions, and situations and which ends in injury, damage, or interruption.
- An accident is the occurrence of a sequence of events that usually produces unintended injury or illness, death, or property damage.
- An accident is an undesired event or sequence of events causing injury, ill health, or property damage.
- An accident is an undesired event that results in harm to people, damage to property, or loss to process.

The contact phase in the accident sequence is traditionally often referred to as the *accident* segment of the sequence, which is incorrect as the accident is the total sequence of events and the loss (injury and damage) is the last phase of the event.

The National Safety Council (USA) defines an accident as: "that occurrence in a sequence of events that usually produces unintended injury or illness, or death and/or property damage." This definition, too, refers to the contact and exchange of energy where the harm is done as the accident phase of the sequence of events. The entire sequence of events, the loss causation sequence, is the accident. The unintended injury referred to is caused by the exchange of energy.

Frank E. Bird, Jr. and George L. Germain, in *Practical Loss Control Leadership* (1996), define an accident as "an undesired event that results in harm to people, damage to property, or loss to process."

In analyzing these definitions, it is clear that the factors leading up to a contact are undesired and the resultant effects, after the contact, are also undesired. A simple yet effective way to distinguish between an accident and a near miss incident is that the accident results in a loss and the near miss incident doesn't.

From here on, this book will describe an accident as "An accident is an undesired event, which results in harm to people, damage to property or loss to process." (Loss is experienced.)

CONFLICTING DEFINITIONS

What causes the most confusion concerning the recognition of a near miss incident are definitions that describe an incident as some event that may or may not have caused injury. That could be anything. The American National Standards Institute, Inc. (ANSI), Standard: ANSI/AIHA Z10–2005, *Occupational Health and Safety Management Systems*, is one of the many definitions that cause this confusion and applying its definition of a near miss will confuse the issue more. The institute defines a near miss as an incident:

> An event in which a work-related injury or illness (regardless of severity) or fatality occurred or could have occurred (commonly referred to as a "close call" or "near miss") (p. 17).

This is totally confusing. No wonder safety personnel are inclined to call all events "incidents." Many refer to accidents/incidents to make sure all events are covered, which is also misleading. Was the event an accident or a near miss incident? This book endeavors to separate accident and near miss incident by clear definitions and descriptions of the two similar, yet very different, events.

ACCIDENTS VERSUS NEAR MISS INCIDENTS

In some accidents, there also can be near miss incidents involved. In a boilermaker workshop, a pressure vessel explodes due to a faulty relief valve. Shrapnel from the exploding boiler (damage) flies across the work area (energy) injuring two employees (accident) and narrowly missing (near miss incident) a group of workers who are working on a nearby milling machine. The shrapnel flying over their heads misses them, constituting a near miss incident. The event injured two employees and damaged property and, therefore, is an accident.

ACCIDENTS, NEAR MISS INCIDENTS, AND INJURIES

For the purpose of the following chapters, it is important to understand the following:

- An accident is an undesired event that actually results in injury, damage, business interruption, or combination thereof.
- A near miss incident does not result in any injury, damage, or business interruption, but has the potential to do so under slightly different circumstances.
- An injury is the resultant physical harm to a person's body (including occupational illness and disease).

DEFINING AN INJURY

An injury is also defined as "the bodily hurt sustained as a result of an accidental contact. This includes any illness or disease arising out of normal employment."

The contact with a source of energy could cause injury to people. The word injury includes occupational illness and disease. The injury is a direct result of contact with a substance or source of energy greater than the resistance of the body. The item that inflicts the injury is the agency that could be an occupational hygiene agency, or a general agency. Injuries caused by accidents are normally immediate (acute). Industrial diseases are mostly long-term (chronic) as they manifest over a period of time. The exchange of energy in diseases is normally referred to as exposures and occurs over a time period. There is an exchange of energy as in an injury accident, except it is phased over a longer time.

General Agencies

General agencies include:

- Walkways
- Machines
- Ladders
- Sharp edges
- Machinery
- Equipment
- Power/hand tools

Occupational Hygiene Agencies

Occupational hygiene agencies are those items that cause the illness or disease. They include:

- Gas
- Heat
- Noise
- Fumes
- Radiation
- Ergonomic defects
- Insufficient lighting
- Chemicals, etc.

DEFINITIONS: INJURIES AND DISEASES

WORK INJURY

"A work injury is any injury suffered by a person, and which arises out of, and during the course of, his normal employment." The definition of work injury includes occupational disease, work related disability, and occupational illness.

Occupational Disease

"An occupational disease is a disease caused by environmental factors, the exposure to which is peculiar to a particular process, trade or occupation, and to which an employee is not normally subjected, or exposed to, outside of, or away from, his normal place of employment."

Injury Compared to Accident

Most people confuse accident and injury. Not all accidents result in injury, and there is a definite distinction between the term *accident* and *injury*.

An accident is the event and an injury is a consequence or end result of the event. The end result may have multiple consequences, such as property or equipment damage, process interruption, etc. The severity of the injury caused by an accidental event is difficult to predetermine, or define. The "luck factors" referred to later explain how the severity is sometimes determined by absolute fortune, either good or bad.

Trying to reduce the severity of the injury is a postcontact safety control. Quick evacuation, prompt medical treatment, adequate medical facilities, and trained personnel all contribute to the reduction of the severity of the injury. The recuperation time after an injury depends on numerous factors. It also determines the number of shifts lost as a result of the accident. These losses, in turn, determine the total costs of the accidents.

FACTS CONCERNING UNDESIRED EVENTS AND NEAR MISS INCIDENTS AND ACCIDENTS

- The majority of undesired events (high risk acts, high risk conditions, and near miss incidents) do not end up in injury. Less than 1 percent of all undesired events result in serious injury (injury-producing accidents), approximately 2 percent result in minor injury, and about 5 percent cause damage to property, material, and the environment. Based on the Bird–Germain (1992) 1:10:30:600 ratio, the majority are ranked as near miss incidents.
- Accidents and near miss incidents are not planned or budgeted for.
- All accidents result in some form of loss, which can be tied to a cost.
- Near miss incidents do not result in a loss.
- Accidents and near miss incidents occur as the result of a sequence of events.
- There is normally more than one cause for an accident and/or near miss incident.
- Fortune, chance, or luck plays a major role in determining the outcome of high-risk acts and high-risk conditions.
- The severity of an injury is also fortuitous.
- The majority of accidents and near miss incidents can be prevented.
- A small percentage of accidents are beyond control due to natural factors.
- Accidents indicate poor management control as a result of a failure to assess the risk.
- Accidents are often described as "a series of small blunders."

- An accident that results in serious injury has possibly occurred previously, but did not culminate in injury.
- Failure to assess the risk and take necessary action is the main cause of preventable accidents.

ACCIDENT SEQUENCE

At this stage, it is pertinent to examine the loss causation model, or accident sequence, and understand the sequence of events that lead to a near miss incident, accident, and subsequent loss.

Accidents are caused by a sequence of events, a combination of circumstances and activities that culminate in loss, similar to a snowball or domino effect. The loss may be an injury, damage, or business interruption. Due to some unexplained circumstance, sometimes called fortuity or luck, the event does not end in loss and this is usually termed a near miss incident. The factors leading up to an accident are there, but the event is interrupted as there is no exchange of energy and, therefore, no injury, property damage, or loss.

FAILURE TO ASSESS THE RISK

The first factor in the loss causation sequence is the failure to assess and mitigate the risk. As Dr. Dan Petersen (1997) said:

> A firm can dictate, in advance, what actions it should take to prevent accidents, and then it can measure how well these predetermined actions are executed (p. 37).

Risk assessment is a method that is predictive and can indicate potential for loss. With this knowledge, an organization is then able to set up the necessary management controls to prevent these risks resulting in losses, such as injuries, property damage, business interruptions, and environmental pollution. This method of accident prevention entails examining near miss incidents, risk assessing, and ranking their potential and investigating and rectifying the root causes of the high-risk, near miss incidents.

Many safety programs focus on the consequence of loss and not the control. Effective risk assessment is proactive, predictive safety in the finest form. In risk assessment, the keywords are: "It's not what happened, but what could have happened."

LACK OF CONTROL

The second link in the accident sequence is lack of control. This lack of safety management control could be no safety program, no safety program standards, or noncompliance to the standards or lack of a structured safety management system. This triggers the basic causes of accidents. If no formal, near miss event reporting and investigation system is in place, this would be classified as an inadequate control system.

BASIC CAUSES OR ROOT CAUSES

The basic (root) causes of accidents are categorized as personal and job factors. They are the underlying reasons why high risk acts are committed and why high risk conditions exist. A personal factor could be a lack of skill, physical or mental incapability to carry out the work, poor attitude, or lack of motivation. Job factors could include inadequate purchasing, poor maintenance, incorrect tools, or inadequate equipment.

These basic causes then trigger the immediate causes that are unsafe work conditions and unsafe work practices (high risk conditions and high risk acts).

Immediate Causes

High Risk Conditions

High risk conditions are physical work conditions that are below accepted standards. This results in a high risk situation or an unsafe work environment.

High risk work conditions include:

- Unguarded machines
- Cluttered walkways
- Poor housekeeping
- Inadequate lighting
- Poor ventilation

Near miss incident control will highlight high risk practices and conditions before they result in an accident.

High Risk Acts

High risk acts are the behaviors of people that put them, and possibly others, at risk (at-risk behaviors). This means that people are behaving contrary to the accepted safe practices and, thus, creating a hazardous situation that could result in a loss.

High risk acts include:

- Working without authority
- Failure to warn somebody
- Not following procedures
- Rendering safety devices inoperative
- Clowning and fooling around in the workplace

Accidents and near miss incidents are always a result of multiple causes, normally a combination of high risk conditions and practices, and seldom, if ever, is an accident or a near miss incident attributable to a single cause.

Natural factors account for a small percentage of accidents. Tornadoes, thunderstorms, volcano eruptions, earthquakes, and floods are examples of natural or environmental factors that can lead to major losses. These can neither be attributed to high risk behavior nor an unsafe work environment. Taking up an unsafe position or tempting the elements would contribute to, or aggravate the severity of, a loss in a natural event, but does not cause the event itself.

CONTACT AND EXCHANGE OF ENERGY

The high risk conditions or acts give rise to an exchange of energy and a contact that is the stage in the accident sequence where a person's body or a piece of equipment is subject to an external force greater than it can withstand, which results in injury or damage.

A luck factor exists here because the high risk act or condition may only result in a near miss incident with no loss. There is no contact with the energy or the energy is insufficient to cause harm. For example:

- In scenario one, a person is driving a motor vehicle and fails to stop at a stop sign at an intersection. This is a high risk act. The action had potential for loss.
- The same person is driving down the road in scenario two and the vehicle fails to stop at the stop sign. This is the high risk act and the car speeds through the intersection. However, another vehicle, which has the right of way, also passes by at the same time, narrowly missing the vehicle that failed to stop. Here we have a flow of energy as the two cars narrowly missed each other, but no contact or exchange of energy took place and there was neither damage nor injury. This is a near miss incident.
- The same event occurs in scenario three, but this time an oncoming car is speeding down the road and there is a collision. The losses are injury to the drivers and damage to the vehicles.

What determined the difference between scenario one and two? The difference was good fortune or luck. This is sometimes attributed to timing or positioning at a certain moment. Some believe it was being at the wrong place at the right time. The driver was lucky that there was not an approaching car and he got away with committing the high risk act of not yielding. In scenario three, the luck factor has proved unfortunate and the perpetrator was unlucky; there was a contact, exchange of energy, an accident took place, and there were losses.

A near miss incident must have an energy phase or there is no near miss scenario. A high risk act or condition does not constitute a near miss incident if there is not a flow of energy that could have contacted. They should be reported and acted upon, anyway.

The energy phase must consist of a flow of energy and not merely potential energy. A suspended load has potential energy, but a worker walking under the load is not involved in a near miss incident. He is committing a high risk act and the suspended load is a high risk condition (no matter how secure the load is).

If the load falls, we have a flow of energy, and if it narrowly misses the employee when it falls, we have a near miss incident. This is if the load was recovered before it fell to the ground completely causing damage.

There could be a flow of energy and an exchange of energy that is also classified as a near miss incident, but only if the exchange of energy was below the threshold limit of the body or structure. The load fell and just scraped past the employee's sleeve causing no injury.

If there was no perceivable damage or injury the event would remain as a near miss incident. Agreed, that there was a loss of time due to the event and the subsequent investigation, etc.

If the falling load damaged the goods being hoisted, or the surface onto which the load fell, we would have an accident scenario rather than a near miss incident scenario because of the substantial loss. Remember that if the goods fell causing damage and narrowly misses the employee, then there is a property damage accident and a near miss incident combined.

Many events result in injury as well as damage and also involve near miss incidents, all outcomes of one event.

INJURY, DAMAGE, OR LOSS

After the contact and exchange of energy, luck again plays a role in determining the outcome of the contact. The outcome could be injury to people, damage to property, harm to the environment, or process interruption, or a combination. We have no control over the outcome of the contact. Once the process is in motion, no control activity whatsoever can determine the outcome, which could be minor injury, serious injury, negligible, or severe damage to property or even death.

INJURY

If the contact results in an injury, we are again dependent on luck. The injury may be minor, disabling, or fatal. The outcome of the exchange of energy and subsequent injury is fortuitous and depends on luck. The end result of a contact cannot be predicted or controlled. Contact safety controls (at the time of the energy exchange), such as personal protective equipment, safety belts, and vehicle air bags, contribute to help reduce the severity of injuries that are hard to predict.

PROPERTY DAMAGE

One of the three major outcomes of a contact is property damage. Accidental property damage is damage caused by an accident, which does not result in injury or business disruption. Many safety programs do not call for the reporting or investigation of these damage accidents, which in most cases also have potential to cause injury to employees under different circumstances.

The damage is usually a result of a contact and exchange of energy greater than the resistance of the object. Property damage can include damage to buildings, floors, equipment, machinery, and material.

In referring to the accident ratios, the property damage accident occurs more often than any other type of accident. Property damage accidents, therefore, are opportunities to identify the basic cause, and take steps to eliminate a similar accident occurring. It will be appreciated that should a similar accident occur, because of hazards that have not been rectified, the outcome may be different, The next time the accident may result in injury, damage, business interruption, or a combination of all three.

Property damage accidents are the most important in the accident ratio. They also are warnings that a failure exists in the management system. This causes root causes to exist, which in turn, give rise to immediate causes and the contact, which then causes a loss in the form of damage to equipment, machinery, etc.

Property damage accidents are often a result of motorized vehicles colliding with the building, the cladding, the raw product, or the finished goods. Most manufacturing concerns are intent on the throughput of the plant and cannot afford the final product to be damaged by accidents. All property damage accidents should be reported and investigated. They should receive the same attention as an accident that causes serious injury.

Most property damage accidents have the potential to injure people, therefore, they should not be ignored. All significant property damage accidents should be thoroughly investigated and a costing done to indicate the actual losses incurred. Costs of repairs to equipment and vehicles should be listed and tabled as well at the various safety committee meetings. These statistics form a vital part of loss statistics.

The environment also can be damaged as a result of fire or pollution. Extensive losses can occur even though no injuries take place. Most property damage does result in business interruption and financial loss.

Fire

Fires are devastating. Every year, millions of dollars worth of property and products are destroyed. Fires are undesired events and occur as a result of high risk acts, high risk conditions. Property damage caused by fires is overwhelming. Instances can occur where the fire causes no injuries, in which case, the only consequence is damage to property, machinery, and products.

Business Interruption

A contact need not necessarily cause injury or damage, but may well end up in some form of interruption of the business at hand.

The interruption may either be major or minor depending on the severity of the contact. Invariably, a contact causes some form of loss. If substantial time is lost restarting a machine or rectifying a continuous process that has been interrupted as a result of an accidental contact, it is a loss. The losses caused by business interruption may not be as severe as losses incurred by injuries or property and equipment damage accidents. The exchange of energy in a business interruption is sometimes not as severe, but is sufficient to disrupt the work.

The work output would be affected because of the delay. Extra effort is needed to rectify this delay. Time to clean up, readjust, to realign, are all losses as a result of the business disruption. In certain instances, a critical part may be affected by the contact and, if not damaged, may be malfunctioning or temporarily displaced. All business work, process, and flow interruptions also cost money.

Loss

Each accident results in some form of loss and all losses cost money. Time may be lost, forms need to be filled out, and the business is interrupted to a degree. Many of the costs of an accident are hidden and, therefore, go unnoticed. Direct costs or insured costs are normally the only costs associated with an accident and are the lesser of the two amounts.

Costs

The final phase of the accident sequence and the last link in the chain reaction are costs. All contacts and exchange of energy result in some form of loss. Losses could include both direct and indirect costs of the accident. In mining and industry, property damage costs could be up to 50 times greater than the direct costs of accidents. A third cost is the totally hidden costs that are seldom identified or tallied. The totally hidden costs of the accident are also losses that are hard to determine, but that exists nevertheless.

Part of the management control function would be costing out the accidental losses and showing these as part of the losses of the business. Well-known management consultants have stated that maximizing profits is not the only aspect of business, as minimizing losses is just as important. *Benefit of RoS*

& near miss

A MEASURE OF SAFETY

Because an injury is minor does not mean that the event that caused the injury was. The event should be investigated and the potential and probability of recurrence evaluated. The next similar event may have more serious consequences as a result of luck factors (under slightly different circumstances).

Most safety programs count the serious injuries as a measure of "safety." This measurement method, while still accepted, is, in fact, a measurement of failure. Assessing and controlling the risks of the business and the activities that make up the control measures should be audited and the result will be a more positive measurement of management work being done to combat loss (safety).

Treating the Symptoms

In the accident ratio proposed by many safety professionals, the serious or fatal injury represents the tip of the triangle or iceberg. One cannot focus on the tip of the iceberg, as the tip is the result of the base of the iceberg, or the underlying cause of the loss. In the accident/near miss incident ratio, the high risk conditions or behaviors, or a combination thereof lead to the accidental exchange of energy that causes loss.

NEMIRR (Near Miss Incident Recognition, Reporting, Risk Ranking, Investigation, and Remedy)

Focusing on the tip of the iceberg, or the serious injuries, is treating the symptoms of the problem and not the cause. Near miss incident recognition, reporting, investigation, and remedial action (NEMIRR) offers management an opportunity to react to accident warnings and to eradicate the problems before they result in loss to people, property, plant, or the environment.

As one safety practitioner put it:

> I believe that there is huge opportunity to reduce actual workplace accidents by ramping up the focus on near misses to the same level as actual accidents.

THE ACCIDENT RATIO

Most people involved with workplace safety are aware of the iceberg theory, the safety triangle, or its correct terminology, the accident ratio. For every recorded injury or loss sitting above the surface, there are many unrecorded near miss incidents submerged below the surface. This was first proposed by H. W. Heinrich in 1931 when he published his 1:29:300 ratios.

Statistics tell us that there could be as many as 600 near misses for each one serious injury. According to the Bureau of Labor Statistics (BLS), 3.3 million injuries occurred during 2009. If we multiply each injury by 600, the result is 1.9 billion near miss incidents, or opportunities to prevent accidents, for 2009 alone.

According to the National Safety Council's *Injury Facts* in 2008 some 26 million accidental injuries were experienced during that year. This all inclusive figure includes work-related, recreational, and home injuries. Using the Bird ratio, this converts to at least 7 billion near miss incidents or warnings that may have preceded these actual injuries.

The accident ratio depicted in Model 1.1 shows that for every serious injury as a result of an accident there are some minor injuries, more property damage events, and plenty of near miss incidents. The only way to reduce the injuries that make up the peak of the triangle is to identify, investigate, and rectify the near miss events before they result in injuries or other losses.

As the late safety entrepreneur Frank E. Bird, Jr. explained to me:

> If you look after the near miss incidents, the accidents will look after themselves. You see, you can't be accident free until you are near miss incident free.

MODEL 1.1   The accident ratio conclusion.

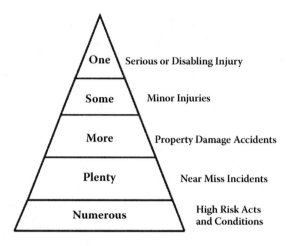

MODEL 1.2   The lower level of the accident ratio.

This ratio which has been developed and researched for a number of years is perhaps one of the most important axioms in safety philosophy. The introduction of thorough near miss incident recognition, reporting, and rectification programs in industry have proved without a doubt that if one reduces the number of near miss incidents the consequent number of injury-producing accidents is reduced considerably.

EXTENDING THE ACCIDENT RATIO

While the near miss incidents that form the base of the accident ratio are truly the foundation of a major injury, numerous high risk acts and conditions lie below on the next level (Model 1.2). Research has indicated that this lower level of unsafe situations could equate to as many as a thousand high risk situations for every serious injury experienced. While the actual numbers are debatable, the fact remains that there must be numerous high risk acts and conditions for the plenty of near miss incidents experienced.

RISK ASSESSMENT

Not all near miss incidents should be investigated. Many organizations think that this is the correct thing to do and find their near miss incident program failing because employees and safety staff are bogged down in investigating every near miss incident reported. While this is humbling and has good intent, it is not the correct approach.

RISK RANKING

Each near miss incident should be risk ranked as to its loss severity potential and probability of recurrence. Only those with high potential should be investigated initially. Once an organization has control over the high potential near miss events, it can then direct its efforts to investigating the lower potential events.

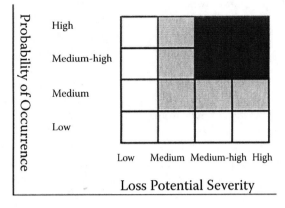

MODEL 1.3   A simple risk matrix.

Based on the safety management principle of the *critical few,* the top 20 percent of near misses could hold the potential for 80 percent of the potential accidental losses.

RISK MATRIX

An invaluable safety tool is the risk matrix (Model 1.3). Remember, it's not what happened, it's what could have happened. The risk matrix is a crystal ball to predict the future or possible outcomes of near miss incidents. It can be used to forecast the probability and severity of the next loss.

High potential events, such as near miss incidents that did not result in loss, should be investigated as rigorously as serious injury-producing accidents, if the assessment of the risk shows in the black (high-high) on the matrix. Risks that rank (high, medium-high) or (medium-high, medium-high) also should be subject to an investigation. All injuries and loss-producing events (in excess of $1,000) should be investigated irrespectively.

NO-BLAME SYSTEM

Reporting all undesired events, such as near miss incidents, is perhaps the most important aspect of any safety program. A no-blame system should be introduced to encourage reporting without consequence. Most near miss incident programs fail as a result of disciplinary steps being taken once an event has been reported. The more warnings that are turned in, the more the opportunity to investigate, identify, and rectify the root cause before a serious loss occurs.

AVIATION

In the U.S. aviation industry, the Aviation Safety Reporting System (ASRS) has been collecting confidential, voluntary reports of close calls (near miss incidents) from pilots, flight attendants, and air traffic controllers since 1976. The system was established after TWA Flight 514 crashed on approach to Dulles

International Airport near Washington, D.C., killing all 85 passengers and 7 crew in 1974.

The investigation that followed found that the pilot misunderstood an ambiguous response from the Dulles air traffic controllers, and that earlier another airline had told its pilots, but not other airline pilots, about a similar near miss incident. Some familiar safety rules, such as turning off electronic devices that can interfere with navigation equipment, are a result of this system. Due to near miss incident observations and other technological improvements, the rate of fatal accidents has dropped about 65 percent, from 1 in nearly 2 million departures in 1997 to 1 fatal accident in about 4.5 million departures in 2007, according to The New York Times.

In the United Kingdom, an aviation near miss incident report is known as an "Airprox" by the Civil Aviation Authority. Since reporting has begun, aircraft near miss incidents continue to decline according to the same source.

FIREFIGHTERS

The rate of firefighter fatalities and injuries in the United States is unchanged for the past 15 years despite improvements in personal protective equipment, apparatus, and a decrease in structure fires. In 2005, the National Firefighter Near Miss Reporting System was established, funded by grants from the U.S. Fire Administration and Fireman's Fund Insurance Company, and endorsed by the International Association of Fire Chiefs and Firefighters. Any member of the fire service community is encouraged to submit a report when he/she is involved in, witnesses, or is told of a near miss event. The report may be anonymous, and is not forwarded to any regulatory agency.

HEALTHCARE

In healthcare, the Association of periOperative Registered Nurses (AORN), a U.S.-based professional nurses organization, has put in effect a voluntary near miss incident reporting system covering medication or transfusion reactions, communication or consent issues, wrong patient or procedures, communication breakdown, or technology malfunctions. An analysis of incidents allows safety alerts to be issued to AORN members.

The Patient Safety Reporting System (PSRS) is a program modeled on the Aviation Safety Reporting System and developed by the U.S. Department of Veterans Affairs (VA) and the National Aeronautics and Space Administration (NASA) to monitor patient safety through voluntary, confidential reports.

The near miss incident registry is a risk free, anonymous reporting tool for near miss incidents in internal medicine. It is sponsored by the New York State Department of Health and administered by the New York chapter of the American College of Physicians. This tool collects information about both near miss incident medical errors and the barriers that kept these errors from reaching patients.

The Railroad Industry

CIRAS (Confidential Incident Reporting and Analysis System) is a confidential reporting system modeled on ASRS and originally developed by the University of Strathclyde for use in the Scottish rail industry.

CONCLUSION

Near miss recognition, reporting, risk ranking, and remedy offers organizations opportunities to rectify potential loss-producing accidents before they happen.

2 The Safety Philosophy behind Near Miss Incidents

INTRODUCTION

According to the National Safety Council (USA), and other safety researchers and pioneers, a large percentage of all accidents are preceded by one or more near miss incidents. In other words, close calls should be wake-up calls for employers and employees to realize that something is wrong in the system and needs to be corrected.

TRACKING NEAR MISS INCIDENTS

Writing for *Occupational Hazards*, S. L. Smith (1994) said:

> Near-miss investigations war with the tradition of using an accident to trigger a thorough look at safety conditions and training. Williams, project manager for safety at Raytheon services in Nevada, suggested that if the purpose of safety programs is to prevent accidents, then tracking near-misses offers organizations a better opportunity to lever their preventive efforts. Near-misses can help employers pinpoint trouble areas and focus their safety efforts and training (p. 34).

Most organizations do not encourage employees to report near miss incidents. Because there has been no injury, little importance is placed on the seemingly trivial event. Most near miss incidents have some form of potential for injury and loss, no matter how trivial they may seem. Although there may not have been a serious outcome, these near miss incidents could result in future accidents.

NEAR MISS—OR NEAR HIT?

Many argue that the event should be called a near hit as this will get management's attention quicker than near miss. Both descriptions are true as: "If it didn't miss, it would have hit." The American Society of Safety Engineers (ASSE) refers to near miss in their dictionary of technical terms, so this will be the preferred term in this book.

BENEFITS

The main benefit of a near miss recognition, reporting, investigation, and remedy (NEMIRR) system is the fact that by recognizing near miss incidents and taking action to correct the underlying problems, an organization will not only reduce the number of near miss incidents, but, more importantly, will reduce the number of actual accidents in the future. Reducing the number of near miss incidents will fix the problems before they can cause accidents. Another major benefit of near miss reporting is that it is easier to get to the root causes of the event because nobody has been injured or killed, which means there is no pressing need for a cover-up.

Near miss incidents also can be defined as: "close calls that have the potential for injury or property loss." Most accidents can be predicted by close calls. These are accidents that almost happened or possibly did happen, but simply didn't result in an injury this time around. In fact, all the stages of the accident were present in the correct sequence except for the exchange of the energy segment that would have caused the injury, damage, loss, or a combination thereof.

EXAMPLES OF NEAR MISS INCIDENTS

Below are some true-to-life examples of actual near miss incident reports. The reports have been edited to make for easier reading, but the contents have not been changed.

- An employee tripped over an extension cord that lay across the floor, but avoided a fall by grabbing the corner of a desk.
- An outward-opening door nearly hit a worker who jumped back just in time to avoid a collision with the door.
- Instead of using a ladder, an employee put a box on top of a drum, lost balance, and stumbled to the ground. Although the employee was shaken, there was no injury.
- A pry bar that was left on the bottom of a mill under repair flew through the air like a missile when a loosened liner fell onto it while the mill was being rotated. The heavy bar missed workers in the immediate vicinity and neither injury nor damage was sustained.
- An employee tested the brakes at the beginning of the shift and they checked out okay. As he approached another vehicle, he hit the brakes and they did not work and he narrowly missed the other vehicle.
- While walking from his car to the stairs, an employee was almost struck by a fast-moving pickup.
- A miner was pulling out hoses to set up the jack leg and the hoses hung up making the miner mad. He pulled really hard and lost his balance and fell down.
- After changing the engine oil on a vehicle, the employee left the discharge gun in the fill tube of the tank and the air purged in the line and spat oil out the tube nearly striking him in the eyes.
- Two welders received minor shocks when they touched a welding machine. On inspection, a loose wire was discovered in the plug of an extension lead that was connected to the welding machine.
- While the employee was in the process of lowering a side plate of a bulkhead, the safety catch on the hook opened and the grab that held the plate almost unhooked.

- An employee was walking toward his job on the demarcated walkway inside the big workshop. The forklift driver rode toward the workshop. At the corner, the forklift driver nearly knocked the employee over because he was unaware that the forklift was approaching.

RED FLAGS

When near miss incidents like these happen, most workers are simply relieved they were not injured and then forget about what happened moments later. However, when employees narrowly avoid injury like this, they may have just been lucky (the three "luck factors"). Because there was no injury, the event is not reported, therefore, the accident causes will not be identified or rectified.

Another person is very likely to be injured by that very same hazard or practice in the future. In fact, the difference between a near miss incident and a serious injury (contact or no contact with a source of energy) is often a fraction of an inch or a split second of time. This can be defined as being lucky. These are red flags waving at employees to let them know something is unsafe and requires immediate attention.

A GIFT

As one safety practitioner put it:

> Near miss incident reports are a gift. They allow an organization to analyze potential accidents, such as fatal accidents, collisions, fires, and explosions, enabling them to take corrective action to improve and rectify processes and procedures before loss-producing accidents happen.

Safety is traditionally taken seriously, but only for a short period of time after an accident. Proactive safety is when serious attempts are made to predict when the next accident may happen by identifying and reducing risks and fixing things before the event may occur. Near miss incidents offer such an opportunity and will involve a change in safety culture.

PRECURSORS TO ACCIDENTS

Near miss incidents are the precursors to accidents that lead to an exchange of energy and subsequent loss. Literally, fractions of an inch or a split second may be the difference between a serious loss-producing accident and a near miss incident. For example, an individual may be on a ladder and the ladder starts slipping sideways, but fortunately catches on a protruding nail or bolt. This should be reported and investigated so that ladder safety could be addressed throughout the organization and the need for tying off ladders emphasized.

As mentioned by Phimister, Bier, and Kunreuther (2004) in "Appendix D: A Note on Definitions," they define the use of the term accident precursors:

Near miss and its analogs, near hit and close call, are other terms that are likely to arise frequently during workshop discussions. Although near misses are clearly related to precursors, we have tried to distinguish them from precursors, and we encourage you not to use them interchangeably. One way to define a near miss (or, equivalently, a near hit or close call) is as an almost complete progression of events—a progression that, if one other event had occurred, would have resulted in an accident. A near miss might consist of one or more precursors that did occur, and one that did not. A near miss can be considered a particularly severe precursor (p. 198).

HEED THE WARNINGS

Six operators at an aluminum casting wheel were waiting for a crucible of molten aluminum to be lowered to the pouring table when the heat from the pot melted the grease on the crane cable causing the break to slip and the full crucible slammed down onto the platform splashing 80 pounds of molten metal all over the area normally occupied by the operators. However, this time the operators were standing on the opposite side of the wheel and none were injured by the flying molten metal. Although this was a property loss accident, the six workers were involved in a near miss incident with high potential. When interviewed, not one of the operators could explain why, on that particular day, they took up a different position. Normally, they would have been standing in the path of the metal splash.

In another instance eight workers were 30 feet off the floor on a scaffold erected in an aluminum smelter pot line. A crane operator proceeded to move the overhead crane from one side of the building to the other. This particular gantry had two cranes mounted on it with different lifting capacities, which shared the common overhead gantry. As he was moving the crane, his focus was on the main 50-ton hook as it traversed the length of the building. Meanwhile, the smaller hook on the 10-ton crane was dangling halfway down at the end of the gantry and unnoticed by the operator. This hook snagged the edge of the 30-foot scaffold and slowly but surely started to lift it off of the ground, tilting it alarmingly. Fortunately, the workers screamed and shouted at the operator who managed to stop the crane movement before the scaffold was completely overturned.

HIGH POTENTIAL FOR LOSS

The next near miss incident not only indicates the luck factor, but also the high potential for loss. An underground miner related this event:

> We were trying to open the man-way doors to go to the surface. We could not open the door because the door to the shaft and the door to the station were open causing a temporary vacuum. We finally got the door open and the pressure of the air pulled me in toward the shaft. I almost ended up falling down the mine shaft had it not been for the retainer wire.

Boylston (1990) refers to near miss incidents as potential problems. He also quotes the luck factors by saying that failure by an organization to recognize, evaluate, and implement controls for early warnings of potential problems usually results in a system of reactive approaches.

> Consequently, there is little if any way to control the magnitude of the problem. Such organizations are "lucky" or "unlucky," depending on the situation. This is no way to manage an organization (p. 103).

FACTS ABOUT NEAR MISS INCIDENTS

- Many near miss incidents have the potential to cause injury to people.
- Near miss incidents are largely ignored because they do not result in any loss.
- Most near miss incidents aren't investigated because a loss did not occur.
- They have often been called near hits, near misses, or near accidents, as they are warnings of potential accidents.
- All high potential, near miss incidents should be reported and investigated to determine and rectify root causes.
- Luck plays a major factor in determining whether the events become accidents (contact) or near miss incidents (no contact).
- Near miss incidents are caused by system failures and can be prevented.
- Reporting, investigating, and eliminating near miss incidents will lead to a reduction in property damage and injury causing accidents.
- Multiple causes are also evident in near miss incidents.

Near miss incidents are warnings and, if an organization eliminates the near miss incidents, the accidents causing damage and injury will look after themselves.

CONTACT (ENERGY EXCHANGE) TYPES

In an accident, there is some form of contact or energy exchange that is above the threshold limit of the body or structure and that causes the damage or injury. In near miss incidents, there is no exchange of energy. There is a flow of energy that misses. If there is a contact, the amount of energy exchanged is below the threshold limit and no loss is incurred.

BUSINESS INTERRUPTION

A contact and exchange of some form of energy need not necessarily cause injury or damage, but may well end up in some form of interruption of the business at hand. The interruption may either be major or minor depending on the severity of the contact. Invariably, a contact causes some form of loss and, if substantial time is lost restarting a machine or rectifying a continuous process that has been interrupted as a result of accidental contact, it is a loss.

The losses caused by business interruption may not be as severe as losses incurred by property and equipment damage accidents or injuries. The exchange of energy in a business interruption is sometimes not as severe as that which caused damage to the equipment, machinery, and environment, but nevertheless is sufficient to disrupt the work.

The work output would be affected because of the delay. Extra effort is needed to rectify this delay. Time to clean up, time to readjust, or time to realign are all losses as a result of the business disruption. In certain instances, a critical part may be affected by the contact and if not damaged may be malfunctioning, or temporarily displaced. All business work, process, and flow interruptions also cost money.

THE ACCIDENT RATIOS

Over the years, researchers around the world have investigated the near miss theory and have compiled numerous accident ratios. They have researched the ratio between the near miss incidents, accidents causing damage, minor injuries, and serious injuries. Most of the research has indicated that there are more near miss incidents that have no visible sign of loss than injury- or damage-producing accidents.

HEINRICH ACCIDENT RATIO

One of the first researchers was H. W. Heinrich (1931) whose accident ratio showed that for every 330 accidents (undesired events) there were 300 accidents that caused no injury, 29 that caused minor injury, and only 1 that caused a serious or major injury. The conclusion was that for every injury-causing accident there were numerous other undesired events (near miss incidents) that had the potential to cause injury (Model 2.1).

When this research was done in the 1930s, the term *near miss incidents* was not yet used and these were referred to as accidents (with no injury). These 300 noninjury accidents could be termed near miss incidents if no damage or interruption occurred. This led industrialists to think about the noninjury accidents and give attention to them.

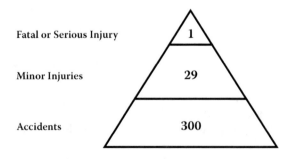

Fatal or Serious Injury   1

Minor Injuries   29

Accidents   300

MODEL 2.1 The Heinrich accident ratio.

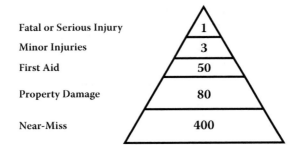

Fatal or Serious Injury	1
Minor Injuries	3
First Aid	50
Property Damage	80
Near-Miss	400

MODEL 2.2   The Pearson accident ratio. (From the British Safety Council. 1974/1975 *Tye-Pearson theory*. With permission.)

TYE-PEARSON ACCIDENT RATIO

In 1974–1975, the Tye-Pearson theory was conducted on behalf of the British Safety Council and was based on a study of almost 1 million accidents in Britain (Model 2.2). The ratio showed that for every 1 serious injury experienced, 3 minor injuries occurred, 50 first aid injuries took place, 80 accidents caused damage, and there were in excess of 400 near miss incidents. The study was concluded by stating that:

> There are a great many more near miss incidents than injury—or damage—producing ones, but little is generally known about these.

FRANK E. BIRD, JR. AND GEORGE GERMAIN ACCIDENT RATIO

In 1966, Frank E. Bird, Jr., and George Germain compiled an accident ratio. This accident ratio study changed the concept by proposing that for every 641 events, there were 600 near miss incidents with no visible loss, 30 property damage events, 10 events resulted in minor injury, and 1 serious or major injury was experienced. This analysis was made of nearly 2 million accidents reported by approximately 300 participating companies employing 1.7 million employees. The Bird and Germain report involved 4,000 hours of confidential interviews by trained supervisors. Their report and subsequent ratio highlighted the occurrence of near miss incidents that under slightly different circumstances could have resulted in injury or property damage.

In referring to the 1:10:30:600 ratio, it should be remembered that this represents accidents and near miss incidents reported and not the total number of accidents or incidents that actually occurred.

THE HEALTH AND SAFETY EXECUTIVE (HSE) ACCIDENT RATIO

This accident ratio, which was derived from a study by the Health and Safety Executive of Great Britain in 1993 (Model 2.3), showed that for every serious or disabling injury, 11 minor injuries were experienced and 441 accidents occurred that damaged property.

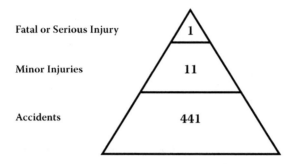

Fatal or Serious Injury — 1

Minor Injuries — 11

Accidents — 441

MODEL 2.3 The HSE accident ratio. (From Health and Safety Executive, U.K. 2006. Online at: www. hse.gov.uk/statistics/causinj/index.htm

SOUTH AFRICAN RATIO

The author (Ron McKinnon), in *Safety Management* magazine (September 1992), elaborated:

> A ratio such as this is truly invaluable, since it makes prediction possible. As soon as too many near miss incidents are reported in a specific area, it is a clear sign to the management and safety staff that a more serious accident is about to occur and they can take preventive action (p. 12).

South Africa's National Occupational Safety Association (NOSA) decided that statistics, which will enable it to construct a ratio, would be of great value to industry:

> We want to determine whether there is a unique ratio in this country (South Africa) or whether Bird's Ratio of 1:10:30:600 was accurate (p. 12).

Often the only difference between a near miss incident and a serious accident is generally just a matter of luck. Take, for example, a brick falling off a platform and narrowly missing the head of a passerby. The accident sequence occurred the minute the brick started to fall. The fact that it missed the passerby was not due to any safety intervention, but was purely fortuitous. While this is recorded as a near miss incident, the implications are very serious.

THE ACCIDENT RATIO CONCLUSION

The accident ratio conclusion is a generalization based on the work of safety research and studies conducted in some seven different countries. No figures are used, but the ratio concludes that there are plenty of near miss incidents where nothing happens, but where something might have happened, if circumstances had been slightly different.

In summary, the accident ratio conclusion shows that for every serious or disabling injury an organization has, it could be experiencing some minor injuries, more property damage accidents, and plenty of near miss incidents.

MODEL 2.4   Accident ratio conclusion.

All of these studies clearly show the futility of investigating only the few serious or disabling injuries when there are hundreds of near miss incidents which, if investigated and their causes corrected, would have prevented the occurrence of the more serious injury-causing accidents.

In conclusion:

- There are more undesired events with no consequence than those with consequences.
- Plenty of near miss incidents occur where nothing happens, but where something might have happened, if circumstances had been slightly different.
- This is where safety management control should be concentrated (Model 2.4).

Krause (1997) refers to the accident ratio and explains:

An at-risk behavior whose outcome lies off of the triangle is a near miss incident. An identical behavior whose result lies on the triangle is an accident. Behavior based safety investigators are not confused by the difference in chance outcomes (p. 293).

PREVENTATIVE OPPORTUNITIES

By taking action to reduce the base of the accident ratio triangle, you are aiming to prevent the serious injury accidents at the peak of the ratio triangle from occurring. Hence, the near miss incidents at the base of the ratio are often referred to as "preventative opportunities." This ratio between near misses and accidents often becomes obvious during accident investigations. While interviewing witnesses to an accident, it becomes apparent that similar events have frequently happened before. Only, in the past, fortune has smiled upon the participant and prevented a serious injury from occurring.

INJURIES VIS-À-VIS NEAR MISSES

By constantly eroding the base of the iceberg or accident ratio triangle, successes will eventually start to impact the top of the triangle, i.e., the minor and serious injuries. An interesting example of this was a study carried out in 1998. A near

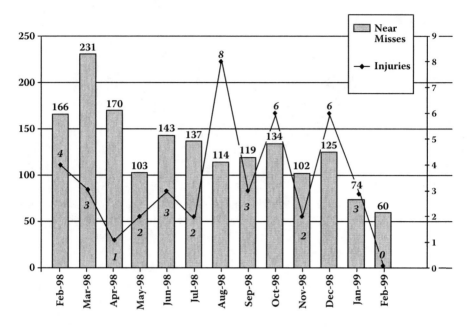

MODEL 2.5 Reduction in reported injuries.

miss incident reporting system was encouraged and from a mere 50 near miss incidents reported in a month, this figure jumped to 231 in the month of March. In superimposing the reduction of injuries (all classifications) on the number of near misses reported, it was found that as the near miss incidents reported increased, the number of injuries fell. The injuries reduced from four in February to one injury in the month of April (Model 2.5). As the number of near miss incidents reported fell, the injury rate once again continued to climb. When only 114 were reported, 8 injuries were experienced.

As is stated in NOSA's MBO Five Star Safety and Health Management System Introduction Booklet (1991):

> The work of any good manager should be to reduce the "plenty" of near-miss accidents (p. 7).

IMMEDIATE ACCIDENT CAUSES

The immediate causes of the contact stage of the accident are the high risk acts and high risk conditions. These are often referred to as the unsafe acts or unsafe conditions.

- A high risk/unsafe act is the behavior or activity of a person who deviates from normal safe procedure.
- A high risk/unsafe condition is a hazard or an unsafe mechanical or physical environment.

Many years ago, research of some 75,000 accidents indicated that the majority of accidents were caused by the high risk acts of people and the minority by the high risk physical, mechanical, or environmental conditions. However, looking back at the investigation techniques used, it is clear that the most obvious cause, human failure, was selected more often because of the "fault-finding" approach to accident investigation. This approach still features prominently today with many thinking that 100 percent of all accidents are the result of unsafe human behavior.

TRADITIONAL RESEARCH

Traditional research proposed by W. H. Heinrich showed that ±88 percent (the majority of all accidents) could be caused by the unsafe or high risk behavior of people. The high risk mechanical, physical, or environmental conditions could cause ±10 percent of all accidents (minority of all accidents) and there is a small percentage (±2 percent) of all accidents that are beyond our normal control and that can be contributed to natural causes, acts of providence, or other phenomena that we can neither predict nor control.

Years of experience and international accident investigations clearly show that all accidents have multiple causes and cannot simply be explained away as worker failure.

HIGH RISK ACTS

High risk acts include:

- Operating without authority, failure to secure, or warn
- Operating or working at an unsafe speed
- Making safety devices inoperative
- Using unsafe equipment or equipment unsafely
- Unsafe loading, placing, mixing, combining, etc.
- Taking unsafe position or posture
- Working on moving or dangerous equipment
- Distracting, teasing, abusing, and startling (horseplay)
- Failure to wear safe attire or personal protective devices

HIGH RISK CONDITIONS

There are numerous high risk work conditions that have a higher level of risk than other work conditions. These vary from workplace to workplace.

COMBINATION OF HIGH RISK ACTS AND CONDITIONS

Numerous accidents are a result of a combination of high risk acts and high risk conditions (National Safety Council, 2010). Very seldom does one isolated act or condition ever result in an accident. Multiple causes are nearly always present in the accident sequence.

In the United States, machinery is one of the top four sources of compensable work injuries and accounts for nearly 10 percent of all injuries. The other sources are manual handling accidents, which cause 23 percent of all injuries; falls, 20 percent; and struck by falling or moving objects, 14 percent.

Although the high risk acts or conditions may be the most prominent factor that cause the accident, the identification and remedy must not stop there. A thorough investigation must be done to determine why the high risk act took place or why the high risk condition exists. This query will inevitably identify the root causes that must be eradicated.

For example, A worker was walking on the factory floor when he stepped in and slipped on a puddle of oil. Upon investigation, it was obvious that the accident and consequent injury was caused by the oil on the walkway, which rendered the floor slippery and unsafe.

Wiping up the oil is a good way of preventing a recurrence of the same type of accident, but will not solve the problem. After wiping up the oil, one may discover another patch of oil farther down the walkway. Cleaning that up will also only be rectifying immediate causes and not the basic reason for the high risk condition.

An investigation will possibly reveal that a fork truck has a leaking engine oil seal and each time it stands with the engine idling a puddle of oil is left on the floor. An in-depth investigation may discover that the wrong oil seal was fitted. By replacing the oil seal, the root cause of oil on the floor is determined. Immediate causes can be misleading and investigations must take place to determine the root cause of the accident.

LUCK FACTORS 1, 2, 3, AND THEIR NEAR MISS INCIDENT RELATIONSHIP

LUCK FACTOR 1

Once a high risk act has been committed, the outcome (or result) of this act depends largely on chance, good or bad fortune, or luck. This is termed *Luck Factor 1*. A high risk condition is a hazard and can result in a number of outcomes depending on Luck Factor 1.

WARNINGS

S. L. Smith (1994) says that:

> If enough near-misses occur, the question is not, will an actual accident ever happen, but when will it happen (p. 33).

Many call the near miss incidents *warnings* and quite rightly so. Safety management is perhaps the only management science where numerous warnings are given before an accidental contact and loss occurs. The only reason that a contact occurs, or does not occur, is because of the luck factor. The warnings should be heeded nevertheless. Many accidents occur because of missed warning signals.

REAL LIFE EXAMPLE

In investigating an accident in which a worker climbed a noncompacted, poorly supported electric utility pole and was killed when both he and the pole fell to the ground, it seemed a clear-cut case that the man had committed a high risk act. Thorough investigation and some five days of gathering further evidence showed that a fellow worker had been doing exactly the same as the victim except on a different pole. This pole was on the same line and was half a mile distant from the fatality site. Investigation showed that this pole was also not completely compacted or supported. The worker had ascended the pole on numerous occasions and had completed the running of the conductors through the isolators. When trying to analyze why the one pole fell, killing its climber, and the other pole did not, one can only conclude that the electrician on the distant pole was lucky. The practice of working on unsupported poles was commonplace and a condoned practice.

Luck Factor 1

Luck Factor 1 (Model 2.6) determines whether the high risk act or condition results in a near miss incident or accident. A near miss incident is an undesired event that under slightly different circumstances could have resulted in injury, damage, or process loss. The slightly different circumstances are largely at the discretion of Luck Factor 1. The difference between a contact "accident" and the near miss incident, "no contact" is a matter of chance. The high risk act or high risk condition, the undesired event, and the slightly different circumstances are reliant on the Luck Factor 1.

Most near miss incidents have the potential to either injure people, damage property and equipment, or interrupt the business process.

Numerous high risk acts are committed daily, but do not result in a contact of any sort. These are not near miss incidents as there has been no flow of energy that, in a contact situation, would have caused injury, damage, or other loss. Many confuse the high risk act and the high risk condition with near miss incidents. To fall into the latter category, there must be a flow of energy.

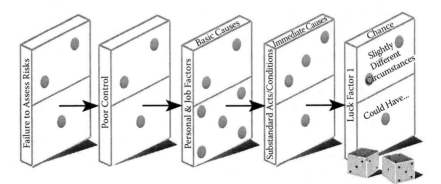

MODEL 2.6   The first luck factor.

High risk conditions may exist for years, but because of circumstances (luck), never result in a contact or any form of loss. Should something happen, a near miss incident, no loss or an accident with loss could occur. The difference is determined largely by chance or by luck. Thus, it can be deduced that the outcome of a hazardous situation is largely fortuitous. A high risk act may result in a contact with resultant loss or may remain an incident that had the potential, but did not cause any loss.

Luck Factor 2

Once an inadvertent and unplanned exchange of energy takes place, the outcome is unpredictable and could result in either injury to persons, damage to property and equipment, or some form of business disruption. Pollution and other forms of loss also may occur as a result of this energy transfer. The results of a contact are unpredictable and the outcomes are normally as a result of chance or Luck Factor 2.

For example, imagine a person walking on a construction site below an unguarded scaffold when a brick is accidentally bumped off and it falls to the ground. In the first instance, the brick falls to the ground and causes neither damage nor injury, but minor process interruption, as the brick has to be picked up and returned to its original position.

In the second case, the same undesired event occurs and the brick falls, this time breaking and creating a loss in the form of damage to material.

In case three, the exact same undesired event takes place, the brick falls from the scaffold and this time hits a worker who happens to be passing by below. The exchange of energy causes minor injury.

In all the three examples given above, there was an exchange of energy, yet the three outcomes or losses were totally different. This is as a result of Luck Factor 2.

It is extremely difficult, and in some cases impossible, to determine the outcome of an undesired exchange of energy. One factor that is inevitable is that the exchange of energy will result in some form of loss, the degree of which is largely determined by chance.

Luck Factor 3

Heinrich, Petersen, and Roos (1969) compiled 10 axioms of industrial safety, the most pertinent one to this chapter being axiom number 4, which states:

> The severity of an injury is largely fortuitous—the occurrence of the accident that results in injury is largely preventable (p. 21).

This fourth axiom is perhaps the most significant statement in the safety management profession. What Heinrich, Petersen, and Roos are explaining is that the degree of injury depends on luck, but that the accident can be prevented. What they further indicate by this axiom is that while the accident can be prevented, the severity is something over which we have little or no control.

In examining the Cause, Effect, and Control of Accidental Loss (CECAL) sequence, once an injury occurs as a result of an exchange of energy, the degree of injury is largely dependent on Luck Factor 3. Most safety activities are focused around the severity of an injury or illness. Consequently, the focus is on an end result, which is determined by fortune, chance, or luck.

Even in the fourth edition of *Industrial Accident Prevention* (1959) H. W. Heinrich and E. R. Granniss further explain Luck Factor 3 by stating:

> There are certain types of accidents, of course, where the probability of serious injury may vary in accordance with circumstances (p. 28).

In other words, the circumstances that determine the probability of a serious injury are really those of either good or bad fortune.

PATTERN

All loss-causation events follow the CECAL pattern, but their progress through the loss-causation sequence is channeled either by Luck Factor 1, Luck Factor 2, or Luck Factor 3. The difference between a fatality, permanent or disabling injury, temporary disabling injury, lost-time injury, and a first aid case is largely a matter of luck.

In examining hundreds of cases where the degree of injury could have been far greater than it was, it is difficult to explain the resultant injury other than by conceding it was luck, as depicted by Luck Factor 3.

Dan Petersen (1997) says:

> Quit looking at accident-based measurements to assess systems effectiveness (p. 40).

What Petersen is referring to is that, because of Luck Factor 3, any measurement of safety performance based on degree of consequence is based on degree of luck. Admittedly, the number of serious injuries and fatalities is important, as is the number of first aid and dressing cases. So much emphasis is placed on lost time, disabling injuries, and reportable injuries that this degree of harm to the body has become the focus of most safety programs, safety practitioners, unions, and regulatory bodies.

EXCHANGE OF ENERGY AND CONTACT

The high risk conditions and acts give rise to a contact that is the segment of the undesired event where a person's body or a piece of equipment is subject to an external force greater than it can withstand, resulting in injury or damage. This is the portion of the accident sequence that is missing in a near miss incident.

A luck factor exists here because the high risk act may only result in a near miss incident with no loss. For example, a motorist fails to stop at a stop sign. This is a high risk act, but there was no loss. The action only had potential for loss. The same action is committed, but this time another car, which has the right of way, narrowly misses the vehicle that failed to stop. There was a flow of energy, but no contact or collision, therefore, no loss. This is an example of a high potential near miss incident.

The same event occurs, but this time an oncoming car is speeding down the road and there is a collision. The losses are injury to the drivers and damage to the vehicles.

In this case, the luck factor has proved unfortunate and the perpetrator was unlucky; there was a contact and an accident and subsequent loss took place.

INJURY, DAMAGE, OR LOSS

After the contact, luck again plays a role in determining the outcome of the contact. The outcome could be injury to people, damage to property, or process interruption. We have no control over the outcome of the contact. Once the accident sequence is set in motion, no control activity whatsoever can determine the outcome.

INJURY

If the contact results in an injury, the severity is dependent on luck. The injury may be minor, disabling, or fatal. The outcome of the injury is fortuitous and depends on luck. The end result of a contact cannot be predicted or controlled.

MEASUREMENT OF SAFETY

The HSE of the United Kingdom conducted an extensive study, which is summarized as follows:

> Any simple measurement of performance in terms of accident (injury) frequency rates or accident/incident rate is not seen as a reliable guide to the safety performance of an undertaking. The report finds there is no clear correlation between such measurements and the work conditions, in injury potential, or the severity of injuries that have occurred. A need exists for more accurate measurements so that a better assessment can be made of efforts to control foreseeable losses.

COSTS

The final phase of the accident sequence and the last link in the chain reaction is costs. All contacts result in some form of loss. Losses could include both direct and indirect costs of the accident. Model 2.7 shows the iceberg effect where the property damage costs could be 60 to 100 times greater than the direct costs. The totally hidden costs of the accident also are losses that are hard to determine, but which exists nevertheless.

COST STATISTICS

Once all the loss-producing statistics have been compiled, the costs should be tabulated and presented. The cost of accidental losses is seldom tallied by organizations that consider them as cost of doing business. Cost can be reduced by effective safety management systems that help reduce expensive property damage and injury accidents.

The costs of medical attention, compensation, and rehabilitation for injuries and occupational diseases should be tabulated monthly and on a 12-month progressive

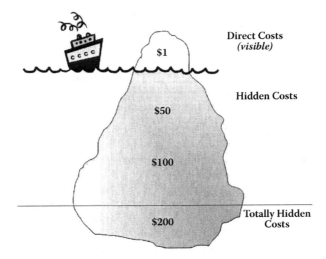

MODEL 2.7   The iceberg effect.

basis. In some instances, it may be necessary to go back in history and record the costs of these injuries and illnesses over a period of years in an effort to determine a trend.

These costs will include the cost of first aid rendered on the premises and also should include the direct and indirect costs.

Direct Costs

The direct costs of any injury-producing accident are normally the hospitalization, the rehabilitation costs, and compensation costs, where applicable.

Hidden Costs

As with all downgrading incidents, the hidden costs are often not visible and add up to more than the direct costs.

Damage Costs

The accident ratio has proved beyond a doubt that there are always more accidents that end up in property damage than end up in injury.

The number of damage accidents should be recorded and tabulated monthly and on a 12-month progressive basis. The damage costing statistic has two aspects and these, like the injury accident, are the direct and indirect costs.

Fire

Fire losses also should be tabulated and reported, both monthly and progressively. Fire losses should include the following components:

- The direct costs of the fire
- The costs of extinguishing the fire
- Indirect costs due to lost orders, delays in production, etc.

If an external fire service was used, the cost of the call-out should be included in the cost of the fires as well.

Production Losses

All the production losses as a result of delays and wastages caused by accidents should be calculated. Although difficult to cost out accurately, an indication or estimated cost will be sufficient to draw attention to the fact that accidents are hampering the production and that there is a loss.

CONCLUSION

Research by many safety authors has clearly shown that the difference in the outcome of an undesired event cannot be predetermined and that in many instances people have been lucky to escape serious injury. The fact that no one was injured is often a cause to ignore what happened and, thus, the potential loss-causing problem does not get fixed. The same event can happen on another occasion, perhaps with different results.

Examining the CECAL accident sequence, it is clear that the three luck factors often determine the difference between a near miss incident and an accident.

Proactive safety involves recognizing, reporting, and eliminating both the immediate and root causes of the event to ensure that we do not rely on luck to prevent injury-producing accidents. A full-fledged NEMIRR system can contribute more to a safety program than most activities based on reactive postaccident actions.

3 Safety Management Functions That Relate to Near Miss Incidents

INTRODUCTION

If the risks arising out of a business have not been identified and assessed, they cannot be managed or controlled. This creates lack of or poor management control, which is depicted by the second domino in the chain of events leading to undesired events.

As Lester A. Hudson (1993) said:

> ... there is a great tendency—human tendency—for management to rationalize after experiencing a human tragedy. It is always so much easier to find the "careless acts" on the part of an injured employee who precipitated the accident, but an enlightened management will not hesitate to look beyond the "unsafe act" on the part of an employee and to consider it as a symptom of lack of management control (p. 2).

MANAGEMENT LEADERSHIP

The safety and health of employees at a workplace is the ultimate responsibility of the management of the organization. Even though it is generally accepted that all share a role in safety, the ultimate accountability lies heavier with all echelons of the leadership. With this in mind, a near miss reporting, recognition, and remedy (NEMIRR) system can only be successful if initiated, led, and supported by line management.

A 1978 NIOSH (U.S. National Institute for Occupational Safety and Health) study identified seven crucial areas needed for safety performance, many of which are not included in most safety programs. The four key elements include:

1. Top management commitment
2. A humanistic approach toward workers
3. One-on-one contact
4. Use of positive reinforcement

The safety charter
Process

In an update, NIOSH promotes the following four safety program elements by stating that an effective occupational safety and health program will include these elements:

1. Management commitment and employee involvement
2. Worksite analysis
3. Hazard prevention and control
4. Safety and health training

Safety charter elements

POSITIVE BEHAVIOR REINFORCEMENT

Positive behavior reinforcement is the key to the success of any near miss incident reporting (NEMIRR) system and, of all the functions carried out by leadership, is likely to have the most effect on the success of the system. It demands playing the ball and not the man. It requires managers to fix the problem and not the person. It forces leaders to deviate from traditional management styles when dealing with issues normally calling for disciplinary measures. It will challenge leadership at all levels, but will help create more positive leadership across the organization.

WHAT IS A MANAGER?

A manager is anyone who uses management skills or holds the organizational title of "manager." A manager is a person who gets things done through other people.

Wikipedia (online dictionary) also gives the following definition:

Management in all business areas and organizational activities are the acts of getting people together to accomplish desired goals and objectives efficiently and effectively. Management comprises planning, organizing, staffing, leading or directing, and controlling an organization (a group of one or more people or entities) or effort for the purpose of accomplishing a goal.

Resourcing encompasses the deployment and manipulation of human resources, financial resources, technological resources, and natural resources.

Management also can refer to the person or people who perform the act(s) of management.

BASIC MANAGEMENT FUNCTIONS

Over the years it has generally been accepted that a manager's main functions include:

- Planning
- Organizing
- Leading or directing
- Controlling

All of these functions entail the management of employees, materials, machinery, and processes, but they do not necessarily include the *safety management* aspect of safety.

Few, if any, functions give specific reference to the activities of a manager pertaining to the safety of his/her resources, namely, people, material, equipment, and the work environment (Model 3.1).

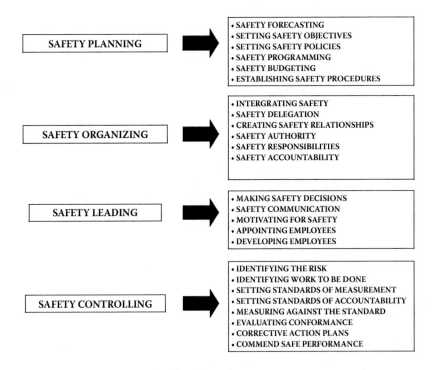

MODEL 3.1   The four basic safety functions of management.

The four basic functions of management have been adapted for safety management and, if integrated into a manager's normal functions, could provide for better management, leadership, and involvement in the safety program and its elements.

SAFETY PLANNING

Safety planning is what a manager does to predetermine the consequences of accidents and potential results of near miss incidents and to determine action to be taken to prevent downgrading events occurring. A main function under this heading is the recording, analyzing, and, if high potential exists, investigation and remedying near miss incident causes.

The Functions of Safety Planning

Safety Forecasting

Safety forecasting is the activity a manager carries out to estimate the probability, frequency, and severity of accidents and near miss incidents that may occur in a future time span. This is usually done by means of risk assessment, critical task identification, and task risk assessment. Near miss incident analysis can predict what losses could have occurred if the event had not been identified and the root causes eliminated.

Setting Safety Objectives

Setting safety objectives is when a manager determines what safety results he/she desires. This would include incorporating a formal near miss incident program (NEMIRR) into the safety system. One objective for the organization could be the reporting of one near miss incident per employee per month. Another objective, for example, would be for at least 90 percent of all near miss incidents to be actioned and rectified each month.

Setting Safety Policies

Setting safety policies is when a manager develops standing safety decisions applicable to repetitive problems that may affect the safety of the organization. Policies and procedures concerning the reporting of near miss incidents, anonymity of reporting, investigations, and feedback would be incorporated into the standard or policy. This written standard would form one of the elements of the organization's safety management system (SMS).

Safety Programming

Safety programming is establishing the priority and following the order of the safety action steps that must be taken to reach the safety objective. An example would be determining what percentage of reported near miss incidents, high risk acts and conditions, and other hazards must be rectified within certain time limits. Also important is setting standards for communicating this information down to the workforce. High potential near miss incidents should be prioritized for investigation. Remedial actions also should be planned in relation to the nature of the hazard to be eliminated.

Safety Scheduling

Safety scheduling is when a manager establishes time frames for the safety program steps. In introducing and maintaining a NEMIRR system, a schedule would be determined for the introduction phase of the system, the training phase as well as the follow up and review of the results, successes, and failures of the system.

Safety Budgeting

Safety budgeting is allocating financial and other resources necessary to achieve the safety objectives. A budget allocation may be required for the reporting incentive scheme. Funds should be allocated. Mechanical or structural repairs or modifications may be needed to eliminate hazards reported through the near miss incident system and these expenses must be budgeted for as well.

Establishing Safety Procedures

Establishing correct safety procedures is when a manager analyzes certain tasks and writes safe work procedures for performing the work. Based on near miss incidents, the jobs can be risk-ranked and the critical tasks identified. This will help prioritize the writing of procedures.

The NEMIRR system needs to be introduced with a procedure for training employees and management in the recognition and understanding of near miss incidents. A reporting modus operandi, i.e., the necessary forms or electronic reporting methods as well as the steps needed to report a near miss incident, must be created. Risk-ranking procedures also are developed along with report back and tracking systems.

Specific investigation systems and methods must be developed and designated investigators trained and appointed.

Safety Organizing

A function of a manager in safety organizing is to arrange for work to be done by the right staff and to be done most effectively. This also would involve allocating people to coordinate the near miss incident reporting system, to do the follow-ups, and produce the necessary reports for management. The manager also appoints investigators and departments to carry out remedial actions.

Integrating Safety into the Organization

Integrating safety into the organization is the responsibility a manager has to allocate safety work to be performed by the various levels within the organizational structure, including the responsibility to recognize and report near miss incidents. The NEMIRR system should be integrated into the day-to-day business of the organization and should not be regarded as a stand-alone item divorced from normal operations. The more the safety management system (SMS) is integrated into the organization, the better the safety culture becomes.

Safety Delegation

Safety delegation is what a manager does to entrust safety responsibility and give safety authority to his subordinates while at the same time creating accountability for safety achievements. All employees are responsible for reporting near miss incidents. Once the root causes of high potential near miss incidents have been identified, managers are then held accountable to rectify the deviations depending on their area of control and level of authority and responsibility.

Creating Safety Relationships

Creating safety relationships is done by a manager to ensure that safety work is carried out by the team with utmost cooperation and interaction amongst team members. The NEMIRR system must be owned by all levels in the organization and should not be seen as a bargaining tool or a system to gain personal benefits or demands. The system requires participation, support, and action from all levels within the organization and cannot be left as one person's or one department's or one manager's responsibility.

Safety Responsibility

Safety responsibility is the safety function allocated to a post. It is the duty and function demanded by the position within the organization. This lies with all levels of management as well as with employees. The higher the management position, the higher the degree of safety responsibility. One cannot be held accountable for something over which one has no authority. The degree of safety accountability also is apportioned to the degree of safety authority. Job descriptions are vital management tools and should clearly define the safety authority, responsibility, and accountability for all jobs and all levels within the organization. The NEMIRR system's safety standard must clearly define these relationships for the system to be a success.

Safety Authority

Safety authority is the total influence, rights, and ability of the position to command and demand safety. Management has ultimate safety authority, therefore, is the only echelon that can effectively implement and maintain an effective near miss incident system. Leadership has the authority to demand the reporting and investigation of near miss incidents and also the authority to take necessary remedial actions to prevent recurrences of the event.

Safety Accountability

Safety accountability is when a manager is under obligation to ensure that safety responsibility and authority is used to achieve both safety and legal safety standards. Employees, too, have safety accountabilities, but in proportion to their safety authority.

Leadership has the accountability to manage a NEMIRR system and to provide the necessary infrastructure and training to enable the system to work. Employees should be held accountable for participating in the system and reporting near miss incidents.

Management at all levels is then held accountable to rectify the problems identified by the near miss investigations and to ensure that the high risk acts or conditions highlighted by the system are rectified and do not recur.

SAFETY LEADING

Safety leading is what a manager does to ensure that people act and work in a safe manner. It entails taking the lead in safety matters, making safety decisions, and always setting the safety example. This is one of the most important management functions in implementing and maintaining a NEMIRR system.

I once accompanied the general manager of a large mine on a safety inspection that the Safety Department did on a regular basis as part of our safety program. After a few minutes of walking, we stopped where there was an obstruction caused by excessive materials that had been dumped in the walkway. He immediately took his near miss booklet out of his pocket and filled in the pocket-size near miss report form, noting the unsafe condition and the fact that employees were forced to step over the obstruction in the walkway. It was great to see senior management participating in the near miss reporting scheme, but it was also one of my most embarrassing moments because I had failed to bring my pocket book with me. Safety leading means doing what you say you are going to do.

The Functions of Safety Directing (Leading)

Making Safety Decisions

Making safety decisions is when a manager makes a decision based on safety facts presented to him or her. Based on the philosophy of near misses and their importance in industry and mines, a manager should make the decision to implement and support such a system.

Safety Communicating

Safety communicating is what a manager does to give and get understanding on safety matters. Management and employees' expectations concerning participation in the NEMIRR system must be clearly communicated. Standards must be set and communicated to all concerning the requirements of their role in the NEMIRR process.

Feedback on reported near miss incidents can either make or break the momentum of the system and is an essential part of such a system. Employees will be skeptical of any new management safety innovation and will constantly test the system. Feedback is essential and a formal communication system ensures that employees are constantly informed of the recording and progress of their near miss incident report. For example, one employee reported a near miss incident when he almost tripped over a depression in the blacktop surface of the employee parking lot. The recommended remedial action listed on the form by the reporting employee read as follows: "The entire car park needs resurfacing." This expensive undertaking was not done and as time dragged on the employee became more and more skeptical about management's sincerity around the near miss program. Eventually, he convinced himself and others that the system was not working and was only lip service.

Because of poor communication, it had not been explained to the employee that even a minor injury would not justify the cost of resurfacing the entire parking lot. Sometimes a cost benefit analysis indicates that a risk, such as the uneven areas of the parking lot, must be tolerated as the cost of eliminating this relatively minor hazard outweighs the cost of a minor injury. I realize this is a rash statement, but it is very true. Cost benefit analysis must be done in some instances as an organization must determine what risks are in the As Low As Reasonably Practicable (ALARP) zone, and which can be tolerated.

Employees are sometimes emotional about safety issues and can be adamant on demanding a "risk free" work environment. This is neither always possible nor practicable and management must communicate this fact to them in such a way that they understand the concept of safety being "acceptable levels of risk."

The point of this true case study is that communication is vital in the NEMIRR process and the very success of the system is dependant on good communication.

Motivating for Safety

Motivating for safety is the function a manager performs to lead, encourage, and enthuse employees to take action for safety. Acknowledging the reporting of near miss incidents by employees is vital to the success of the system and this feedback acts as a reward for such reporting. The power of this small gesture of thanks and recognition cannot be overstated. Incentives may be given for near miss reporting, but these should not be in the form of hard cash or other elaborate prizes. Company T-shirts or coffee mugs are excellent handouts to encourage near miss incident reporting. A mere recognition by the reporting employee's supervisor or manager normally helps motivate continuous reporting. *Avoid any scheme that pays cash money for anything related to safety.*

Appointing Employees

Appointing employees is a management function where management ensures that the person is both mentally and physically capable of safely carrying out the work for that position. The facilitators of the NEMIRR system will be safety and health personnel and management's function in this regard is to appoint the right people in the right positions. The success of coordinating the NEMIRR system will indirectly depend on the support it receives from the safety department and this depends on the quality of safety and health staff initially appointed.

GETMAC

The function of all safety departments and personnel should be to guide, educate, train, and motivate all levels of management, workers, and unions in the techniques of accident and disease prevention and to advise and coordinate the safety system (GETMAC). This should be a staff function and not a line function. All employees have some safety responsibility, but management (senior and line) have ultimate safety authority and, therefore, is ultimately accountable. The safety department cannot, and should not, be held accountable for the safety performance of an organization. This has been stated in numerous instances by a number of safety authors, yet safety departments traditionally drift back to managing the safety of others.

Policeman

Some safety professionals take it upon themselves to give direct instructions to employees. I have known safety practitioners who do inspections and actually give direct orders to workers concerning the wearing of personal protective equipment. In some instances, employees are told by safety staff to clean up their workplace because of poor housekeeping. Employees are often instructed to carry out tasks differently by the safety department. This is line management's responsibility and these instructions should come from direct supervisors not the safety department.

The safety department is not a police force. They should identify deviations from standards and hazards and notify the supervisor. It is the supervisor's job to give instructions to his or her employees. The safety department should not take on this role. If the safety department wants to be policemen, then they should join the security division, and walk around giving instructions and correcting behaviors, etc. Being safety policemen, they belittle the profession of safety practitioners.

When appointing safety and health department employees, management has a duty to the profession to ensure suitably qualified and experienced practitioners are selected. Proper job descriptions based on the American National Standards Institute (ANSI) guidelines: ANSI/ASSE Z590.2-2003 *Criteria for Establishing the Scope* *and Functions of the Professional Safety Position*s, should be used as selection and training criteria for safety staff.

Professionalism

Safety departments should be very professional. Their true place is advising management and coordinating the activities of an ongoing safety management system. They cannot improve the safety by accepting the responsibility for safety. They should not directly try to influence behavior of employees. Only management can do that. Traditional safety is a thing of the past. New approaches are needed by the safety staff if practicing safety is to reduce the risk to people at work.

Developing Employees

A manager develops people by helping them improve their safety knowledge, skills, and attitudes. Management has to ensure the safety and health staff is up to date with the latest trends in safety and risk management and that there is an ongoing self-development program in place for them. Further studies, as well as membership in local, regional, and national safety and health associations, also should be encouraged.

SAFETY CONTROLLING

Safety controlling is the management function of identifying what must be done for safety, inspecting to verify completion of work, evaluating, and following up with safety actions. This is the most important safety management function and has seven steps (ISSMECC):

1. Identification of the risk and safety work to be done.
2. Set standards of performance measurement.
3. Set standards of accountability.
4. Measure against the standard.
5. Evaluation of conformance.
6. Corrective action.
7. Commendation.

1. Identification of the Risk and Safety Work to Be Done

Based on risk assessments, a manager lists and schedules the work needed to be done to create a safe and healthy work environment and to eliminate high risk acts of people. This would mean the introduction of a suitable structured SMS based on world's best practice. All safety management systems should be based on the nature of the business and be risk-based, management-led, and audit-driven. A NEMIRR process should be one element of the structured system.

Safety Management System (SMS)

A safety management system is defined as "ongoing activities and efforts directed to control accidental losses by monitoring critical safety elements on an ongoing basis."

The monitoring includes the promotion, improvement, and auditing of the critical elements regularly.

ANSI

A good guideline for what elements a safety management system should contain is the American National Standards Institute (ANSI), ANSI/AIHA Z10–2005—*Occupational Health and Safety Management Systems*. A similar standard exists for the construction and demolition industry: ANSI/ASSE A10.38-2000 (R2007) *Basic Elements of an Employer's Program to Provide a Safe and Healthful Work Environment*.

BSI

Other international guidelines and safety accreditation systems include the British Standards Institute (BSI), Occupational Health and Safety Assessment Systems (OHSAS 18,001).

Critical Safety Program Elements

The safety management system should contain some 70 key areas that need to be controlled. These are critical elements, and one of those elements should be the near miss system (NEMIRR). According to NIOSH, an effective program includes provisions for systematic identification, evaluation, and prevention or control of hazards, and which goes beyond specific requirements of the law to address all hazards.

What are Critical Elements?

Critical safety elements are elements that include environmental and employee factors and which need to be controlled constantly to prevent accidental losses from occurring.

Critical safety elements are those elements most likely to give rise to losses. Past experience based on thousands of safety inspections and audits have shown that control over certain aspects of the workplace and work practices can significantly reduce near miss incidents and accidents. Many of these critical elements are legal requirements of safety and health legislation.

Controlling critical safety elements is precontact control. It is effort directed toward the prevention of undesired events. Once controls over certain critical elements are exercised, proactive safety is practiced.

Controlling the stacking and storage procedures in the workplace is an example of control over critical safety elements. Ongoing housekeeping campaigns, inspections, audits, and work permits are other examples. A safety management system is designed to exercise control over potential areas of loss. All critical safety elements are potential areas of loss. The NEMIRR system also would be regarded as a critical element because it identifies safety problem areas before a loss occurs and offers an opportunity to fix root accident causes before the event.

Principle of the Critical Few

The principle of the critical few states that "a small number of basic causes could give rise to the majority of safety problems." A few critical jobs could be responsible for the majority of accidents and injuries occurring and these few critical items (critical safety elements) should receive maximum safety control to minimize their potential for causing (the majority) of problems.

Precontact control is directing the safety efforts toward controlling these crucial areas before a loss occurs. Most safety programs are reactive and only institute controls after a loss has occurred. This is termed *postcontact control*: firefighting, patch prevention, or treating the symptom and not the cause.

Why These Elements?

Experience has shown that there are between 60 and 80 critical safety activities (elements) that must be controlled to constitute an effective safety program. These elements may vary from organization to organization and from industry to industry. The emphasis on individual elements also will vary according to the nature of the process, culture of the workforce, and category of business, such as mining, the iron and steel industry, transportation, the fishing industry, manufacturing, etc. There, however, are many common safety system elements, near miss recognition, reporting, investigation, and remedy (NEMIRR) being one.

Benefit

The benefit of controlling critical safety elements is that the work being done to manage safety is channeled at reducing the risk and potential loss in areas that have been identified as crucial. Some critical safety elements help control the physical conditions and the health and safety of the environment, which would contribute to the reduction of losses as a result of an unsafe work environment.

Other critical safety elements are directed toward the control of the persons within a workplace. These controls would include items such as critical task procedures, rules, training, and activities to involve, motivate, guide, and train employees in safe work practices.

Numerous safety elements help control both the behavior of people at work as well as the work environment and the work procedures. No hard and fast dividing line can be drawn between elements defining them as either behavior or environmental control. The one influences the other.

Elements to control the environment include items such as housekeeping, lighting, electrical safety, stacking and storage, ventilation, lifting gear safety, demarcation, machine guarding, hazardous substance control, etc.

Elements to control behaviors and to ensure safe work procedures could include critical task procedures, rules and regulations, appointment of health and safety representatives, holding of regular meetings, safety communication, safety promotion campaigns, medical examinations, etc.

2. Set Standards of Performance Measurement

Standards in safety are referred to as *measurable management performances*. Standards are set for the level of work to be done to maintain a safe and healthy environment free from actual and potential accidental loss. Standards are established in writing for all the safety and health management system elements. Without standards, the management program has no direction nor are safety expectations established. (If you don't know where you're going, any road will take you there.)

Standards for the NEMIRR System

Standards will entail what the system will comprise of, who will be involved, what infrastructure will be needed, what the reporting expectations are, etc. Setting standards of measurement establishes what must be done concerning all aspects of the near miss incident system.

An example of a standard for a NEMIRR system is:

1. Objective

   To define the methodology for reporting and investigating noninjury (loss producing) accidents and near misses so that the immediate, and basic (root) causes of the events are identified and recommendations to prevent a recurrence are proposed and implemented.

2. References

   Applicable legal references and Company Safety Element 4.2 Near Miss/ Accident Investigation.

3. Definitions

   Accident: An undesired event that causes harm (injury or ill health) to people, damage to property, or loss to the process (production and/or business interruption). This includes fires, as there is a loss.

   Near Miss: An undesired event, which, under slightly different circumstances, could have caused harm (injury or ill health) to people, damage to property, or loss to the process (potential production or business interruption). There is no loss.

   Serious Accident: Fire, landslide, or explosion resulting in losses to production or to production equipment (as defined by local legislation).

   Responsible Person (RP): This is the Manager/Superintendent/Supervisor of the area in which the event occurred, or the most senior person on the site where the event occurred.

   Investigation Form (NMAIF): The Company Near Miss/Accident Investigation Form is the form to be used for the investigation of all high potential near misses, property damage, and injury producing accidents.

   SHE: Safety, Health, and Environment.

4. Standard

   • All accidents that result in injury or damage, and all near misses, shall be reported promptly so that an investigation can be launched to determine the root causes, so that corrective action can be taken to prevent recurrence.

   • The Near Miss Accident Initial Report Form is to be posted on the e-mail system before the end of the shift by the responsible person (RP) in the department that experienced the event.

   • A risk assessment of the event must be indicated on the risk matrix included in the form.

   • The company Near Miss/Accident Investigation Form (NMAIF) is to be used for the investigation of all high potential (use risk matrix on initial report) near misses, property damage, and injury producing accidents.

- For all accidents involving cars/pickups/trucks on company property, in addition to the abovementioned forms, a separate form, the Company Vehicle Accident/Damage Report Form must be completed by Company Security because without this form outside repairs cannot be done.
- Accidents that have high potential for loss should be thoroughly investigated.
- Results of investigations shall be shared with others via five-minute talks, SHE committee meetings, and similar avenues, as appropriate.

5. Procedure and Responsibilities (property damage/environmental/vehicle accidents/near miss incident).

Any employee involved in, or who witnesses, a damage accident, near miss incident, however trivial, shall:

- Notify his supervisor immediately.
- Arrange to make the scene of the accident safe and ensure that site evidence is not destroyed unless unavoidable to prevent further injury or damage.
- Cooperate with the investigation.

The responsible person (RP) upon being notified of the property damage or near miss shall (where applicable):

- Visit the scene of the event.
- Arrange to make the scene of the accident safe and ensure that site evidence is not destroyed unless unavoidable to prevent further damage or injury.
- Obtain written statements from the witnesses as soon after the event as possible, but preferably during the same shift.
- Publish the Near Miss/Accident Initial Report before the end of the shift.
- Commence the accident/near miss investigation process and head the investigation meeting.
- Complete page 1 of the near miss/accident investigation form within 24 hours and submit to the SHE Department.
- Finalize the investigation as soon as possible and implement the remedial measures immediately.
- Circulate the investigation findings to all sections within the department and the rest of the plant for their information.
- Ensure that both the immediate and root causes of the event are identified.
- Follow up to ensure that the remedial measures have been implemented as soon as is practicable after the event.
- Ensure that the investigation form is completed correctly and submitted to the next level of authority (one-up manager) for signature.
- Investigate any serious accident as defined by law. ... (Insert local legal requirements.)

6. Investigation and Reporting requirements
   - Near misses and damage accidents involving company activities
   - High potential near miss incidents and accidents causing damage shall be reported and thoroughly investigated using the NMAI Form.
   - The responsible person (RP) will nominate an Investigator for the particular event (Nominated Accident Investigator [NAI]) and commence the investigation with him. The NAI should not always be the SHE superintendent.
   - The RP also will initiate and circulate the Near Miss/Accident Initial Report within the same shift, even if digital pictures are not available in time.
   - The front page of the NMAI form is to be completed and a copy sent to the SHE Department within 24 hours by the RP.
   - The safety superintendent (SAS) (SHE Dept.) will record the event in the register and allocate a tracking number for follow-up purposes.
   - During normal working hours, the investigation shall be started as soon as possible after the occurrence, and completed within 72 hours. If more time is required, notify the central register holder (SAS).
   - Outside of normal working hours the investigation shall be initiated by the RP in whose area of responsibility the event occurred and a NAI shall be nominated.
   - The NAI shall record findings on the Near Miss/Accident Investigation Report Form and both the immediate and basic causes of the event must be determined. The estimated costs of the losses should be entered on the form.
   - The investigation shall include recommendations for actions to prevent recurrence listing:
     - WHAT should be done to prevent a recurrence.
     - WHO is responsible for doing the work/taking action.
     - WHEN the actions are to be completed.
   - Only once these actions have been implemented, should the NAI and RP and the SHE superintendent sign the form.
   - The signed form must now be circulated to the next level of supervision for comments and signature.
   - Managers are accountable for ensuring that the immediate and root causes of the near miss/accident have been identified and that they have been eliminated or mitigated and should only sign the form once this has been done.
   - Once signed by the manager, all high potential near misses, damage/interruption investigations are to be circulated to the appropriate general manager for comments and signature.
   - Once signed by the general manager, the form is returned to the SHE Department for filing and sign-off in the register.
   - The completed investigation report shall be issued within 72 hours of the event.

- Advice and assistance in investigation of high potential near misses and property damage cases can be obtained from the Departmental SHE superintendent or SHE Department.

Near Misses/Accidents Involving Contractors

- The Company person responsible for contractors (to whom the contractors report) in each area (i.e., directly supervised or those on a fixed contract) shall ensure that appropriate reporting and investigation of near misses and damage/interruption accidents is carried out according to this standard as if the contractor was a company employee.

3. Set Standards of Accountability

Management now set standards of accountability by delegating authority to certain positions for ongoing safety work to be done. Coordination and management of the NEMIRR system needs to be allocated to certain departments and individuals and this standard dictates who must do what, and by when, to run the system. The preceding example of a standard clearly defines who is responsible for a number of aspects of the near miss incident reporting and investigation system.

4. Measure against the Standard

By carrying out safety inspections, the actual condition of the workplace and the ongoing activities of employees are now measured against the accepted safety standards. Physical inspection of the workplace will highlight instances of property damage accidents and will trigger a follow-up on the control systems to ascertain if there has been a report and consequent investigation.

What gets measured gets done, and if there is no formal system of measurement, then management does not know how well the NEMIRR system is doing compared to its own standards and best practice.

5. Evaluation of Conformance

Depending on which measurement method is used, the results are now quantified in the form of a percentage allocated, marks given, or a ranking established. Safety audits, both internal and external, evaluate compliance with an organization's standards and scores then indicate whether there is a deviation from the prescribed standards set.

6. Corrective Action

The amount of corrective action will be proportional to the amount of deviation from the standards set. Corrective action may involve enforcing the safety standards and taking the necessary action to regulate and improve the methods.

Corrective actions are the defining stages in a near miss incident system and are where the rubber meets the road. The near miss incident investigation will indicate what immediate and root causes triggered the event and the only way to prevent a recurrence of the event that may result in an accident is to take corrective, preventative actions. This may mean fixing the high risk condition or correcting high risk behavior or a combination of both, and then rectifying their root causes.

Once again, standards are established for these corrective actions and they state who-must-do-what-by-when in order to get the situation rectified. Corrective action must be positive, time related, and be assigned to responsible people.

7. Commendation

Commendation is when a manager pays a compliment and expresses gratitude for adherence to an achievement of preset safety standards. This is applicable across the board and also is pertinent to the NEMIRR system. If employees are not recognized for participating in the system, their enthusiasm will soon wane. If the system is introduced as another "flavor of the month," employees will not integrate near miss incident reporting into their daily routines, but rather treat it as another safety fad that will soon die a natural death.

Recognizing and reporting near miss incidents must become a part of the company's culture and management's involvement, support, and participation in the system is vital. This integration of the NEMIRR system into the day-to-day activities of employees at all levels will drive the NEMIRR system into becoming a part of the safety culture of the organization. Managers and employees alike should carry their reporting booklets or forms at all times and also report on positive items and actions noticed. Reporting near miss incidents will become a natural thing to do. That's when it's clear that there has been a change in the safety culture.

Safe Behavior Recognition

The NEMIRR system also is used to report positive and safe behavior noticed as well as safe working areas. A part of the recognition system is recognition of safe behaviors and conditions. Often management tends to focus on the negative aspects of safety, yet its most powerful tool is to recognize the positive. In safety, positive behavior reinforcement goes a long way in achieving employee participation and involvement in the safety and near miss reporting system.

One of the criteria of a NEMIRR system is that employees' names are not necessarily listed on the report forms so the reports are anonymous and the process does not become a head hunt. Employees appreciate this and feel motivated to take part. Traditionally, they were found to be at fault and discipline most likely played a role in the process. This approach is perhaps the reason for worldwide nonrecognition and nonreporting of near miss incidents by employees. Recognition, on the other hand, will enthuse and encourage workers to take part in the system and lead to its success.

4 Safety Management Principles Relating to Near Miss Incidents

PROFESSIONAL SAFETY MANAGEMENT PRINCIPLES

The safety aspect of management can be summarized as follows:

- Safety management can clearly be identified and classified.
- The precontact, contact, and postcontact safety management control can be measured.
- Safety management has a specific vocabulary.
- Safety management has certain principles and fundamental truths that are derived from professional management, but which are applicable to safety management.

SAFETY MANAGEMENT PRINCIPLES

The management profession has many principles that have developed over the years that are applicable to safety management. When implementing a near miss incident program, the involvement of all must be sought. The application and practicing of these principles can assist in ensuring system success.

The following principles will indicate the importance of management's role in near miss incident management. The near miss incident reporting system (NEMIRR) must have the full support of and must be driven by management at all levels. Management also must participate in reporting near miss incidents and, as leaders, set the example.

A director of a large city division took the lead in implementing a world's best practice safety management system throughout the division. One of the 70 key program elements was that of near miss incident reporting. At a management meeting a few weeks later the chair he was sitting on in the boardroom collapsed and he fell to the floor, fortunately not injuring himself. He promptly filled in a near miss incident report that was circulated to the entire workforce. As a remedial measure, all the boardroom chairs were scrapped and replaced with new, sturdier ones.

Once an employee submits a near miss incident report this will be the test for management to see if a positive, humanistic approach is followed. Most managers at this stage resort to traditional management tactics and want to punish the employee for committing a high risk behavior or for not fixing the high risk condition. Management should rather view the report as an opportunity to fix the problem and not the worker. The near miss reporting system needs to operate in a positive management/worker climate.

Managers should visit the workplace as often as they can and interact with employees at the point where the work is done. This is also where the risks lie and these visits and contacts with employees will help reinforce management's visible commitment to the safety system.

PRINCIPLE OF MANAGEMENT RESULTS

A leader tends to secure most effective results through others by performing the management work of planning, organizing, leading, and controlling.

PRINCIPLE OF SETTING SAFETY OBJECTIVES

The speed, efficiency, and motivation to carry out safety work are increased if the work is directed toward preset safety objectives. This means that management must know what it wants and decide what work is necessary to achieve the safety goal. Safety responsibility must be assigned and the working relationship must be clearly defined in relation to the safety objectives. The NEMIRR system must have a set of standards and targets. Expectations must be set for the number of near miss incidents to be reported per month or year, but care should be taken not to allocate near miss incident quotas to individuals as this may lead to fudging of the books to meet the quota.

PRINCIPLE OF RESISTANCE TO SAFETY CHANGE

The introduction of safety standards and procedures that differ from the way things were done in the past tends to be met with resistance by the people involved. Introducing safety management is a change that creates an insecure environment. Introduction of safety systems require adequate preparation and the best way to introduce a comprehensive safety management system is by introducing it element-by-element. The smaller the changes, the smaller the resistance to the change. In South Africa, we call this "eating the elephant—one small bite at a time."

The introduction of a near miss incident reporting system will normally be met with resistance and skepticism by the workforce. This will be harder to overcome if discipline was levered when near miss incidents were reported in the past. Employees will resist this change because of a number of reasons discussed later in this book.

PRINCIPLE OF SAFETY COMMUNICATION

The more people are informed about the safety requirements and achievements, the more they are motivated to participate and accomplish safety results.

Effective communication improves motivation, and reasons must be given as to why certain steps have been taken for safety, e.g., why we are introducing a near miss incident reporting system. Highlights and safety achievements must be communicated to the workforce.

For the success of the NEMIRR system, reported near miss incidents must be summarized and circulated for organization information. This information feedback is for all employees and management of the site. The publication of these reported near miss incidents can be via the weekly safety newsletter, via internal e-mail, or can be posted on the company's internal Web site and on the safety notice boards.

The information should indicate what happened in the form of a brief description of the situation, the risk ranking of the potential, and the action taken (or to be taken) to rectify the sources of the problem. This will form part of the personal feedback to the employees, which is a vital part of the near miss incident program. People reporting hazards and near miss incidents want to receive acknowledgment of their contribution. This communication is imperative if the system is to succeed.

A question often asked is whether or not the reporter's name should be required on the near miss incident report forms and appear on the subsequent weekly or monthly summary. It may be that general publication of near miss incident reports could create peer pressure if employees realize the near miss incident information was to be widely publicized within the organization. If near miss incidents are widely reported, employees will be exposed to many reports and an additional report should not draw attention to the reporter, thereby minimizing peer pressure.

PRINCIPLE OF SAFETY PARTICIPATION

Safety motivation increases in proportion to the amount of participation of the people involved.

Safety involves all people, and safety activities should involve all people. Employees should be informed of the facts at all times and should be asked to give input and suggestions on aspects of safety that directly or indirectly concern them.

A near miss incident reporting system levels the playing field and employees and managers at all levels can contribute to the system by reporting near miss incidents. Since the attaching of their names to the reports is optional, they are afforded some security in the anonymity of reporting. Many who have never partaken in the safety system events can now become active members without fear of repercussions or reprimand.

PRINCIPLE OF SAFETY DEFINITION

Decisions concerning the safety program can only be made if the basic (root) causes of loss-producing events are clearly identified. "A logical and proper decision can be made only when the basic or real problem is first defined." (Prescription without diagnosis is malpractice.)

Often a lot of manpower is wasted by directing efforts into the rectification of immediate causes of loss-producing events. The principle of safety definition states that the basic or root cause must be identified before a remedy is prescribed.

As can be seen, the close calls or warnings in the form of near miss incidents clearly indicate the basic or root causes of systems' failure and offer the opportunity to rectify the failings before loss occurs. They are the true accident indicators.

PRINCIPLE OF SAFETY AUTHORITY

Participation in the safety program and motivation to accomplish results increases if people are given authority to make decisions concerning safety. Safety objectives must be set and people must know what their authority in the safety management system is. Ownership of a segment of the safety system can help lead to participation and safety success. All employees, irrespective of their standing, should be given the ability (authority) to report near miss incidents, high risk conditions, high risk behavior, or safe work.

PRINCIPLE OF INTEREST IN SAFETY

The workforce will only become interested if management show an interest in the safety results achieved by them individually and as a group. Management must set the safety trend and help them achieve safety objectives. Reporting near miss incidents must be a common interest in the organization.

Taking action on reported near miss incidents will indicate to the workforce that management is indeed taking safety action. Reporting back on the number and nature of near miss incidents reported generates tremendous interest in safety. Most near miss incidents are interesting occurrences in their own right and remedial actions applied also create interest and involvement of the employees.

PRINCIPLE OF SAFETY REPORTING

The higher the level to which safety personnel report, the more management cooperation they are likely to obtain. Safety coordinators are the catalysts in the safety system and should function as such. Safety coordinators should not be in a line function, but rather in an advisory capacity in a staff function.

PRINCIPLE OF THE CRITICAL FEW

A small number of basic causes could give rise to the largest number of safety problems. A few critical jobs could be responsible for the majority of accidents and injuries occurring within an organization. A few near miss incidents will be critical as they hold the most potential for losses under slightly different circumstances. Risk ranking will indicate those near miss incidents with the highest probability for loss, the highest severity of loss (should it occur), and the probability of the event recurring. These critical few events should be focused on and receive the same attention as loss producing events, such as accidents.

These few critical items should receive maximum safety control and attention to minimize their potential for causing the majority of problems.

PRINCIPLE OF SAFETY RECOGNITION

Safety motivation increases as people are given recognition for their contribution to the safety effort. Commending and encouraging people for safe acts goes far in ensuring that those safe acts are repeated. Good safety should be praised and this praise should be made in public where possible. The same is applicable in near miss incident reporting by employees. The more employees are recognized for reporting, the more reports they will submit.

It may be difficult for traditional leadership to accept that, even though the employee may have committed an unsafe act, they get recognition for reporting it. Traditionally, this would raise the discipline, or three-days-off-with-no-pay flag, but if an organization can get employees to admit their mistakes and report them in a safe environment, this will constitute a greater learning for the employee involved, the organization, as well as fellow employees.

PAST SAFETY EXPERIENCE PREDICTS FUTURE EXPERIENCE PRINCIPLE

An organization's past safety experience and efforts tend to predict the safety effort and experience of the future. Attempts to get near miss incidents reported have more than likely failed in the past and, if a direct effort and full commitment is not made, may well fail in the future.

As mentioned before, the introduction and maintenance of a near miss incident reporting and remedy system will lead to a safety culture change within the organization. If one considers the effort, commitment, and change to the way safety is normally managed in a plant, the NEMIRR system will bring about a culture change. It will force leadership to become leaders, to walk the talk, and to create a positive environment where employees can feel free and safe to report what was previously regarded as discipline requiring behavior.

Safety management normally experiences more resistance than any other aspect of an organization's business. The safety culture that is embedded in the organization tends to prevail in the future. Safety attitudes and behaviors experienced in the past tend to be carried over to the future. These can only be changed with a deep, ongoing, and concerted effort to make the near miss incident system work.

PRINCIPLE OF SAFETY APPLICATION

The principle of safety application states that the more the various elements of safety are practiced and applied, the more they are understood and accepted as day-to-day activities.

No safety management principle is more applicable to near miss incident reporting than the principle of application. The more the near miss reporting system is used the more it will be understood, accepted, and practiced.

As with any new or different safety innovation, the introduction of a near miss reporting system needs to be well structured and become an integral part of the culture of an organization. It must be an ongoing program and not just a nonsupported safety gimmick or flash in the pan.

Tenacity around the encouragement of reporting and the sincere follow-ups in remedying the root causes of near miss incidents must be evident and effort must be sustained to keep the program going. Once employees (and others) report more and more near miss incidents and see the problems being addressed, the reporting will become a natural thing to do. Eventually the reporting, ranking, investigation, and rectification of problems will become integrated into the day-to-day work process and will become part of daily activities rather than an add-on safety burden.

PRINCIPLE OF POINT OF CONTROL

The greatest potential for control tends to exist at the point where the action takes place. High risk behaviors and conditions mostly exist where the work takes place and, therefore, this is where the majority of near miss incidents will occur. Management is not always where the work is done, therefore, the employees doing the work are key when it comes to the reporting of deviations, such as high risk work practices, hazardous conditions, and the occurrences of near miss incidents.

Accidents and near miss incidents occur where the work is done and this principle states that the best opportunity for safety control is there as well. Remedial measures to prevent a recurrence of an undesired event are best directed at the point of control where the greatest potential for loss exists.

PRINCIPLE OF MULTIPLE CAUSES

The principle of multiple causes states that "accidents, near miss incidents, and other problems are seldom, if ever, the result of a single cause." This pertains to near miss incident investigation, which in itself is another vitally important criterion of any safety system. If the investigation system is not structured and does not follow the loss causation sequence and determines both the immediate and root causes of the event, the system is basically worthless.

Traditionally, the only reasons for an accident, or near miss incident, investigation was to assess blame and find a guilty party. This will never solve the problem or determine the root causes of downgrading events, and will not fix the real cause of the problem. This is termed prescription without diagnosis. Finding one cause of a near miss incident is totally insufficient because there are always a number of reasons for an action or situation.

Many investigators fear delving deeper into the causes of an event as this may open a can of worms. As mentioned previously, a NEMIRR system cannot work unless management and the organization declare amnesty and create an environment within which employees can report near miss incidents confidently, and where investigations can delve into all the causes, irrespective of what will be uncovered.

As stated by the principle of definition, if the real causes of the near miss incidents are not found, how can real solutions be proposed? If all the contributing factors are not investigated, how can the causes of the near miss incident be identified and rectified?

The following excerpt is from the author's book, *Changing Safety's Paradigms* (2007):

SINGLE CAUSE—NEVER!

Most accident investigations uncover a single cause and label this as the main cause of the accident. Accidents are always caused by more than one factor. The principle of multiple causes states that accidents or loss-producing events are seldom, if ever, the result of a single cause. This means that accidents have more than one cause. When H. W. Heinrich did his research prior to 1929, only one cause of the accidents studied were considered, which led to the findings that 88% of all accidents are caused by unsafe acts. At the same time of this research, the National Safety Council found that if one looked at both the unsafe acts and the unsafe conditions, they were almost on a par.

People feel comfortable when identifying an unsafe act after an accident. This gives them the satisfaction that the employee "messed up." He committed an unsafe act. Examples are: the employee breached the standing safety procedures, the employee decided to work unsafe, the employee failed to consider his safety and the safety of others, the employee has an attitude.

The investigation should continue once an unsafe act is identified. If accidents were investigated correctly, other contributing causes would be sought. What about the physical work environment? Are all the machines guarded? Have there been regular audits of the area? Is a structured safety system in place? The major question I would ask when investigating an accident that only lists unsafe acts is: "Why did the system allow the person to commit these unsafe acts?"

Until accident investigation is used as an effective tool to identify the true causes, it is a waste of time. Poorly investigated accidents and near miss incidents are missed opportunities to take positive steps to prevent recurrences. Every accident or near miss incident has multiple causes. Proactive safety means investigating thoroughly and identifying all the immediate and root causes of the event. Then it involves asking why each unsafe act was committed. It further involves asking why each unsafe condition existed. This is opening a can of worms and delves into the basic causes. The employee who was injured is only a victim of the safety system failure.

As the Columbia Space Shuttle Accident Investigation Board reported, "Causal factors for accidents that result in severe injuries are multiple and complex, and relate to several levels of responsibility" (p. 124).

SAFETY SUCCESS VIS-À-VIS MANAGEMENT LEADERSHIP

The degree of integration of safety principles and standards, and involvement, is directly proportional to the amount of management leadership and commitment. The more encouragement and support management gives employees, the more near miss incidents will be reported.

The chief executive officer (CEO) of any firm is the leader and if he is a leader for safety concerns, others will follow. Top management must always set an example for others to follow. Setting an example for safety is of crucial importance in the safety program. Management must set the standards, take the lead, and be prime examples of safety. I once went out for a plant inspection with the general manager of the mine I was working on. As he left his office, he automatically pocketed his near miss reporting booklet. The reporting on near miss incidents or hazards by senior management had become such a value that he automatically took his pocket-size booklet with him on every mine tour. As mentioned previously, I had not taken mine with me and was setting a poor example for the Safety Department.

Involvement in the safety program by the CEO and other members of management is not sufficient, total commitment to the loss prevention movement is required. Participation at safety functions, during safety inspections, and ongoing safety control activities is important.

Most levels of management are involved in safety on a day-to-day basis. This involvement does not necessarily mean they are contributing enough. Total commitment is what is required of all levels of management, especially the executive.

When asked the difference between commitment and involvement, a world-renowned safety practitioner answered, "The next time you have a plate of bacon and eggs for breakfast, remember that the chicken was involved in the process, but the pig was totally committed."

Does the involvement of the executive level come from being forced to comply to certain legislation or as a result of pressure from the workforce and the unions, or is it as a result of identifying and realizing the benefits of complying to accepted norms? A chief executive and his fellow officers have two choices. They can either give minimum commitment to the safety movement or maximum commitment and support. A successful near miss system needs both initial and ongoing management support as well as buy-in from the unions and employees. Line management must take the lead and be active participants.

The initiative to launch and maintain safety momentum is required from top management, so is the leadership and ongoing support necessary to prevent downgrading accidents that lead to losses. All managers should be visibly committed to the safety drive and this visibility will involve their frequent visits to the shop floor to mingle with the workers and manage the safety at the point of action. The closer to the point of action, the more effective the safety management will be.

Interest in the safety of workers, full support to their safety requirements, and the ongoing supply of resources can only be facilitated from the management level.

People normally pay the most interest to what the CEO wants them to pay attention to. If this happens to be safety, the CEO automatically sets the trend. Executive leadership, commitment, and a desire to improve the work standards and reduce high risk behavior and high risk conditions lead to better safety, production, and higher quality standards. Only executives can set safety objectives that can be cascaded down to lower levels within the organization.

Participating at safety committee level, steering the activities that lead to the achievement of safety objectives, and maintaining the safety momentum are prime, but often neglected, functions of management.

CONCLUSION

Safety management control eliminates the basic causes of accidents and near miss incidents by setting up management systems, which include near miss incident cause rectification, and by delegating safety responsibility and accountability. This system creates a work environment in which accident and near miss event root causes are reduced, consequently reducing the high risk acts and conditions.

As Dr. Mark A. Friend (1997) says:

> Only members of the management team can create or change the environment. (And it is, after all, their job to do so) (p. 34).

5 Near Miss Incidents, Myths and Safety Paradigms

NO INJURY—NO ACCIDENT (NO BLOOD, NO FOUL)

The "no blood, no foul" concept has come to mean that the workplace is safe until there is an accident and a resultant injury. A common paradigm is that if there is no injury there is no accident. As long as an organization goes from day-to-day injury-free, then safety is fine. Nobody is being injured so all appears to be in order. Meanwhile, the organization could be experiencing numerous undesired events that do not result in injury or damage.

An accident is an undesired event that results in some form of loss. This loss could be injuries to people, damage to property or equipment, business interruption, or other forms of loss. The undesired events that cause loss and incur costs are accidents. Near miss incidents do not end up causing injury, therefore, qualitative judgments of safety performance, reached exclusively in terms of injury frequency, are apt to be grossly inaccurate.

If one gauges a breakdown in the safety management system by counting injury losses only, a false impression is created. Therefore, a company that is operating, injury-free, is not necessarily safe. In fact, it could be experiencing near miss incidents, which, under slightly different circumstances, could have resulted in injury. Equipment may be damaged, manufactured products damaged, but no one is injured.

The Health and Safety Executive of Great Britain states that:

> Any simple measurement of performance in terms of accident frequency or incidence rates is not seen as a reliable guide to the safety performance of an undertaking.

It suggests that the efforts to control foreseeable risk be assessed instead.

If a workplace is injury-free, it is not necessarily safe. Behind the scenes, the tip of the iceberg in the form of a serious injury may not yet have appeared, but near miss incidents exist at the base. A safe organization means no deviations in the management system, no poor quality products are produced, and production quotas are met without losing money in other areas, such as injuries, damage, and interruptions.

Because no injury occurred does not necessarily mean there is no potential for injury. Most near miss incidents have the potential to produce injury. The undesired events should be prevented. The prevention of deviation from management standards should be a priority of the business. If not, numerous accidental events take place, but, because there is no injury, the hazard is not recognized. An organization that is not injuring employees does not mean that it does not have a safety problem. Safety is defined as the control of all accidental losses and an absence of injuries is no indication that safety is in order.

WHY INJURIES ARE SEEN AS "SAFETY"

One of the main reasons for the safety myth of no injury—no accident is that safety is perceived as being about injury to people. This is another mindset that hampers safety efforts. When an employee gets injured, safety becomes important. It becomes the talk of the day for a few hours. Employees' minds heal quicker than the injury. Workers will mill around, drinking coffee, discussing injury to a fellow worker or other employee. The next day, it is business as usual. Joe's all right, he is recuperating in the local hospital. It will never happen to me. What about all the near miss incidents, close calls, and other no loss-producing events that, under slightly different circumstances, could have caused injury?

DAMAGE

Property damage accidents exceed the number of injury-producing accidents. According to Frank E. Bird, Jr. and George Germain (1996), for every serious injury there are at least 30 accidents that result in property damage. During their study of nearly 2 million accidents reported to the Insurance Company of North America, they found that 30 property damage accidents were reported for every major injury. They also stated that:

> Property damage accidents cost billions of dollars annually and yet they are frequently misnamed and referred to as near accidents (p. 21).

Their line of thinking recognizes the fact that each property damage situation could have resulted in personal injury. If there was sufficient force to damage equipment, then that same force could have inflicted injury. The injury mindset is a carryover from earlier training and misconceptions that led supervisors to relate the term "accident" only to injury.

ACCIDENT RATIOS

In reviewing the accident ratio, less than 2 percent of all accidents occurring at a factory or mine result in injury. Less than 1 percent of undesired events, according to Bird and Germain's ratio, cause serious injuries, which are known as reportable, lost time, or disabling injuries.

This category of injury is used because this is the degree of injury that is mostly recognized in the safety profession. This injury classification is what safety campaigns, safety records, safety performance measurement, safety comparisons, safety competitions, safety recognition schemes, etc., recognize as being the accepted measurement of safety. If less than 1 percent of serious injuries are caused by accidents, safety efforts should not be focused only on them. More effort should be directed at the minor injury-causing accidents, the property damage accidents, and the numerous near miss incidents. All of these have potential to result in severe injury, yet they are often ignored.

The discussed accident ratios propose the ratio between serious injuries, property damage accidents, and near miss incidents. Irrespective of the actual numbers given in these ratios, the facts remain that there are numerous near miss incidents (warnings) before a serious injury-causing accident is experienced. For every one serious or disabling injury, there are some minor injuries, more property damage accidents, and plenty of incidents where nothing happened, but where something might have happened. Changing the safety myth that, if there is no injury there is no accident, is to heed these warnings before it is too late.

RISK ASSESSMENT

The risk of these near miss events should be assessed, and those with the highest probability, severity, and frequency should be subject to an investigation. Once the root causes are sought and found, remedial measures should be put into action to prevent the deviation from standards recurring. Only by preventing a deviation occurring will the consequence, or the likely consequence, be eliminated. Safety is often reactive because organizations sit back waiting for the consequences to occur before taking action. This is reactive safety.

The apathy that exists in organizations, when it comes to near miss incident recognition, is the false sense of security that "nothing happened." Managers are lulled into a false sense of security because nothing apparently happens. Employees don't report these events because "nothing happened." Again, it's a matter of no blood, no foul.

ICEBERG EFFECT

The tip of the iceberg is visible. The tip of the safety iceberg, however, very seldom does the most harm. It is the base of the iceberg, or the larger hidden portion that creates the havoc. The portion of the iceberg under the waterline is what causes most concern. These are safety warning signals.

Minor injuries, property damage, and near miss incidents constitute the hidden part of the iceberg and are the parts of the undesired events that can wreak havoc with a business system. Safety efforts cannot be successful if they are focused on the tip of the iceberg, the serious injury. Safety, if viewed as the control of total accidental losses, must focus on the base of the iceberg. The base of the iceberg represents the accidents that a company has not yet had. These include near miss incidents, or warnings that consequences could follow, under slightly different circumstances. Remember, luck determines the outcome and end result of an undesired event.

NEAR MISS INCIDENTS AREN'T IMPORTANT: A PARADIGM

The ignoring of near miss incidents or warnings because they appear insignificant is a major safety paradigm. For safety to be effective, near miss incidents must be recognized and acted upon. Near miss incidents have often been called the accidents that have not yet happened. Most organizations are years behind with the recognition of the importance of near miss incidents. Most organizations are just as far behind when it comes to the reporting, evaluation, and rectification of these warning signs. The most important sentence in the entire philosophy of safety management is: "It's not what happened, but what could have happened." This means that each near miss incident could have resulted in something else if it were not for luck. Almost all individuals in every group that I have interviewed have admitted that they have experienced near miss incidents. Most had narrow escapes with no injury. Near miss incidents are understood by hourly employees, middle management, the safety profession, and the executive management. If some action is taken on near miss incidents, the root of the safety problem will be tackled.

INJURY-FREE CULTURE

I have often heard the expression "injury-free culture" and other expressions just as catchy and well intended. Other campaigns state: "We are striving for an accident-free culture." Others boast: "We want to be injury free for a year." And, perhaps the most ambitious of them all: "We will have zero accidents." Even though this sounds good, it is almost impossible without bending the rules and hiding the mistakes. Many organizations create an aura, or climate of fear, so that terrified employees do not report accidents, near miss incidents, or injuries and the company fools itself that it has achieved its injury-free goal. Although the intentions of these well-intended campaigns are good, the following truths must be recognized:

- You cannot be injury-free unless you are accident-free.
- You cannot be accident-free unless you are near miss incident-free.
- You cannot be near miss incident-free unless you have no high risk conditions and no high risk acts are committed.
- You cannot eliminate high risk acts and conditions until you have effective control measures in the form of a structured safety management system.
- Your safety system cannot be effective until you assess and reduce the risks.

The bottom of the safety iceberg represents the numerous near miss incidents that occur before there is an accident that causes property damage, minor or serious injury. Focusing on the top of the iceberg, namely, the injury, is futile. An organization cannot eliminate the serious injury at the top of the safety iceberg until the near miss incidents at the base have been reduced substantially. Reducing the number of near miss incidents reduces the probability and frequency of an injury occurring. It erodes the foundations of major injuries.

NEAR MISS INCIDENTS NOT REPORTED

I once presented a lecture at a regional safety congress in the United States, and asked the audience how many near miss incidents they had reported at their companies during the preceding year. Of an audience of 90 people representing major organizations, only 4 indicated that their organizations had reported near miss incidents. This and other research led me to believe that most organizations were lagging as far as their safety systems were concerned.

The reporting and ranking of near miss incidents is perhaps the most important aspect of any safety system. The near miss incidents with high probability and high potential severity should get the same important treatment as actual injury or damage-inflicting accidents. They warrant a full accident investigation, complete remedial measures, and follow up.

Instituting a structured near miss reporting and rectification system (NEMIRR) can make the biggest advancement in safety in any organization. Studying the monthly printouts of excellent near miss incident reporting systems, I noted that seldom did a month go by without at least two of the near miss incidents being ranked with having potential to have caused fatal injury. This means there is an opportunity to rectify the causes of these near miss incidents before the consequences are experienced. An example of the potential of a seemingly insignificant happening that may have gone unreported under different circumstances is this near miss incident report from an underground miner:

> An employee tested the brakes in the beginning of the shift and it checked out OK. As he approached another vehicle, he hit the brakes and they did not work and he narrowly missed another vehicle.

WARNINGS IN ADVANCE

The advantage of the safety profession is that it is the only profession in the world that gives warnings of imminent danger. What is meant by this, according to Bird and Germain (1996), is that there are 600 warnings before a serious injury occurs.

It should be emphasized that the only difference between a near miss incident and a hit (exchange of energy) is luck. Numerous writers have come to the same conclusion. These include Bird and Germain (1966), H. W. Heinrich (1931), W. E. Tarrants (1980), Petersen (1997), and others. Yet, the safety profession seems to ignore this valuable information. Safety can only be successful if near miss incidents are viewed as the true "safety in the shadows," and are identified, reported, and acted upon.

LUCK FACTORS

If an organization has a low injury rate, one must ask the question, "Is it as a result of control or is it as a result of relying on luck?" If it is as a result of relying on luck, then the safety management system needs to be reviewed.

The luck factor is so prominent in safety and yet so poorly understood. Even working a million man-hours without a disabling injury could be as a result of luck. Not having a fatality for the past 50 years could be as a result of luck. Having a fatality once a year could be bad luck. The point is, unless the risk reduction controls are in place, we are relying on luck, not control, in the form of risk reduction. An organization must make sure that it has at least 70 elements that constitute its safety system. This system, if operating correctly, puts it in control of potential areas of loss.

SPACE SHUTTLE COLUMBIA

According to the Columbia Accident Investigation Board (CAIB), the Space Shuttle Columbia's demise was caused by a chunk of insulating foam, which dislodged from the left bipod ramp of the external fuel tank and struck the leading edge of the left wing, causing a breach of the thermal protection system. This, in turn, allowed superheated air to enter the wing and melt the aluminum structure, which caused the wing to fail until aerodynamic forces caused the breakup of the orbiter.

The CAIB also found that foam had been shed on more than 80 percent of the 79 missions and that debris had caused damage on every Space Shuttle flight. There had been insulation shedding on most missions. The report also asks the question why the shuttle missions continued when they were aware of the foam being shed during ascents. The CAIB further stated that it would seem the longer the shuttle program allowed debris to continue striking the orbiters, the more opportunity existed to detect the serious threat it posed. Were damage accidents and high potential near miss foam strikes ignored here?

CONCLUSION

Because no loss occurred, it does not mean that under slightly different circumstances an injury and other losses may have happened. The absence of injury does not necessarily mean that there was no accident. The myth that just because nothing happened all is in order is clearly busted by the fact that near miss incidents are accidents waiting in the shadows.

6 Safety and Health Policies

INTRODUCTION

Safety and health management systems (programs) must start at the logical point, which is a policy of intent that is initiated, implemented, and followed through by management.

Policies have been defined as *standing decisions, which apply to repetitive questions and problems.* A safety and health policy is standing safety decisions, which apply to repetitive safety problems that may affect the safety of the organization.

A safety and health policy, or safety mission and vision, is the commitment that the management teams make to safety and is also the safety commitment to which the employees agree. It is the guiding document of safety within the organization and a declared intent to maintain a safety and health management system. Most near miss incident recognition, reporting, investigation, and remedy systems (NEMIRR) are driven by this guiding document, the safety and health policy.

Although the system is not specifically mentioned in the overall safety and health policy, in most cases, its function and responsibilities are spelled out in a safety management standard on near miss incident reporting, which is a stand-alone document. A safety standard is defined as a *measurable management performance.* Each element of the safety management system should have a written standard that describes the actions, duties, goals, objectives, and responsibilities for each element.

As part of its recommended safety program, the U.S. National Institute of Occupational Safety and Health (NIOSH) recommends the same action:

- State clearly a worksite safety and health policy.
- Establish and communicate a clear goal and objective for the safety and health program.
- Provide visible top management involvement in implementing the program.

SAFETY AND HEALTH POLICY

Planning

The compiling of a safety and health policy is part of the safety planning function of management and is the starting point of any occupational health and safety system. Where possible, reference to the maintenance of a near miss incident system should be included in the main policy.

ADVANTAGES

The following are a few advantages of a written safety and health policy:

- Gives common points of view concerning occupational health and safety.
- Provides rational rather than erratic safety decisions.
- Helps managers at all levels by giving the employees the advantage of knowing the executive's point of view on safety.
- Is an indication of the organization's safety considerations.
- Safety policies allow for delegation of safety work.
- Efficient and effective teamwork is facilitated by a safety policy.
- Provides guidelines to everybody on how to do the right things concerning safety.
- Commits the executive management, line management and employees to a process of continual safety improvement.

THE ESSENTIALS

Most progressive organizations with low injury experience and high physical and environmental standards normally have a credible safety and health policy that includes a near miss incident system commitment.

SAFETY RULE BOOK

It is advantageous to have a copy of the safety and health policy of the organization reproduced in the safety rule book. This helps during the safety induction process and also serves as an ongoing reminder of the safety culture of the organization.

REQUIREMENTS OF POLICIES

Safety and health policies must be dynamic, be realistic, and the objectives set must be tangible. Both management and worker representation should agree on the policies. This joint policy should be extensively publicized. The safety and health policy should cover all aspects of safety including:

- Injury prevention
- Damage control
- Occupational disease control
- Occupational health
- Near miss incident reporting and investigation
- Environmental protection
- Emergency awareness
- Fire control
- Ongoing improvement strategies, etc.

The policy should be appropriate to the risks arising out of the organization's processes and must include a commitment to continuous improvement of safety and health. The policy should be displayed and also made available to outside interested parties and it should also state the intent of complying with all local relevant safety and health legislation.

Ongoing review and modification of the safety and health policy must take place on a regular basis and this must be committed to in the policy.

COMMITMENT

The safety and health policy should be drawn up with considerable care and once agreed to, must be signed, issued, and publicized. Management and worker representatives normally draft the policy and, in some instances, the management team as well as the worker representation team sign the policy.

POSTED AND DISPLAYED

The safety and health policy should be posted at prominent positions throughout the organization. To give credibility to safety, the policy preferably should be attractively printed and suitably framed. The display positions must be carefully selected and workers must constantly be briefed and reminded of the policy.

The agreed-to safety policy must be practical and achievable. More tangible than intangible objectives should be part of the policy. The policy also should list practical ongoing steps that will ensure compliance to the policy and it must be written in such a way that it is understood by all. Ongoing briefing sessions should be in operation to ensure that everyone understands the gist of the policy.

EXAMPLES AND EXTRACTS

To give the reader some true life examples and extracts from current loss control, safety, health and environmental policies, the following are given.

Example 1

This policy is a brief statement that has backup documentation to explain the details of "an accident-free workplace."

> To have a team-based culture in which our actions continuously demonstrate our commitment to an accident-free work environment and the well-being of each and every member of our family.

Example 2

The safety and health policy means that we are committed to:

- Providing and maintaining a safe and neat working environment for all employees.
- Following an agreed procedure to identify, report, record, and rectify unsafe conditions and practices.

- Ensuring compliance along the guidelines of the safety system.
- Providing safety training at all levels.
- Encouraging employees to develop a safety culture.
- Measuring, by means of a formal evaluation system, the extent to which safety objectives have been met.

This organization's policy commits it to implement and maintain a comprehensive safety and health management system, one of the elements being the near miss incident system.

Example 3

The following Occupational Health and Safety Policy was issued by an Executive Safety Leadership team and reads as follows:

Nothing at Our Company is more important than the health, safety, and well-being of our employees and their families. The Company Safety Leadership Team believes that identifying and reducing risks will prevent all forms of accidental loss. As an organization, we are determined to achieve our vision of reducing the probability of illnesses and injuries to our Company family, visitors, and the communities and environment in which we operate.

We are dedicated to:

- Making risk-based occupational health and safety management a core value that drives performance
- Holding managers accountable for occupational health and safety in all of our facilities
- Providing the practices, tools, and resources via the Company Health and Safety Management System

To achieve our occupational health and safety objectives:

Each employee, regardless of position or title, must take individual responsibility for health and safety. It is the job of each employee to create a work environment that eliminates occupational health and safety hazards. Further, we encourage all employees to be role models and leaders in health and safety at work as well as for their families and their neighbors in our communities.

Our Company is committed to complying with and, where excellent practice would demand, even exceeding applicable occupational health and safety laws wherever we operate. We believe that occupational health and safety laws and regulations can and must be integrated with our effort to produce a world-class research environment.

We will not be satisfied until we have reduced the probability, to the maximum extent practicable, of occupational injuries and illnesses occurring among our family of employees. This is the only acceptable goal, and we are dedicated to achieving it through continuous improvement of our safety management system.

We know we will achieve these results only through each employee's participation in, and attention to, health and safety and dedication to make working safely an integral part of every job we do.

(Signed by) Company Safety Leadership Team

CONCLUSION

Not all the safety and health policies given as examples meet the criteria of naming each aspect of safety and health, such as disease prevention, environmental control, etc. It is generally accepted that when the organizations quote *safety*, they include *all* aspects and do not necessarily list them individually. For a successful near miss system to be incorporated, it is advisable to include it as a special line item in the policy and essential for a written standard to be drafted to guide the system.

Safety and health policies should be short, to the point, and, most importantly, the employees must know the policy. It should be remembered that a policy is only as good as the employees' knowledge thereof. Safety and health policies should be revised regularly to ensure that they are kept updated.

Reproduction of the safety and health policy in the organization's annual report also indicates to the shareholders and directors the importance that the company places on its most vital assets: employees, machinery, materials, and the environment.

7 Near Miss Incident Risk Management and Assessment

INTRODUCTION

Occupational health and safety has evolved in an effort to cope with the complexities of modern industry and the risks that accompany manufacturing, mining, transportation, and other activities.

Much development has been in the field of risk management, which considers probabilities of undesired events occurring and enables action plans to be put in place to prevent and reduce the possible losses that could occur.

This chapter endeavors to simplify the risk management approach and give the reader an overall idea as to the various components of risk management and how they combine to form a comprehensive management strategy.

Risk assessment is a vital part of the near miss incident reporting and remedy system (NEMIRR) process and helps one determine the potential probability, frequency, as well as the potential severity of loss of a near miss incident. It helps predicts what could have happened under slightly different circumstances.

THE RISK MANAGEMENT PROCESS

The following are steps in the risk management process:

- To identify all the pure risks within the business and connected to the operation (hazard identification).
- To do a thorough analysis of the risks taking into consideration the frequency, probability, and severity of consequences (risk assessment).
- To implement the best techniques for risk reduction (risk evaluation).
- To deal with the risk where possible (risk control).
- To monitor and reevaluate on an ongoing basis.

DEFINITIONS OF RISK MANAGEMENT

Risk management is:

- The business science, which is applied to manage the pure risks of the business.
- A comprehensive strategy for dealing with risk.

- A group of techniques for minimizing the adverse effects of pure risks.
- The combining of the functions of planning, organizing, leading, and controlling the activities of an organization so as to minimize the averse affects of accidental losses produced by the risks within the organization.

Near miss reporting and remedy systems, risk management, and risk controls helps predict and reduce the potential aversive affects of accidental loss.

GOALS OF RISK MANAGEMENT

The goals include:

- Taking steps to minimize risks arising from the business.
- Reducing the magnitude of the risks.
- Reducing the frequency of exposure to the risks.
- Dealing with the event should it occur.
- Assisting in recovering from the event should it happen.

DEFINITIONS OF RISK

Risk is:

- Any chance of a loss.
- The probability that injury or damage will occur.
- The likelihood of an undesired event occurring within a time frame or under certain circumstances.

TYPES OF RISKS

There are two major types of risk: *pure risks* and *speculative risks*. Pure risks only offer the prospect of loss and speculative risks offer both a chance of gain, or loss. A speculative risk could be the purchasing of stock or investing money in real estate, etc.

Accidents seldom offer any chance of gain. High risk actions offer chances of gain mainly in effort or time saving. Does this mean that high risk actions are risks taken with the probability of either loss or gain as probable outcomes? In interviewing a group of previously injured employees, the majority had been injured in accidents as a result of cutting corners or trying to save time. If this argument is to be held true, then high risk acts are not pure risks, but speculative risks. What makes the difference is the cost benefit analysis. Does the saving of a few minutes warrant the possibility of a crippling injury?

RISK MANAGEMENT COMPONENTS

The two main components of risk management are *physical risk management* and *risk financing*. As indicated in Model 7.1, physical risk management consists of risk

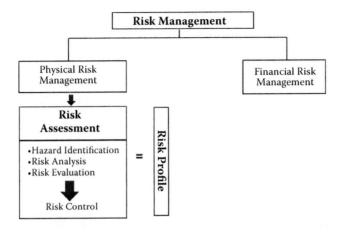

MODEL 7.1   Risk management.

assessment and risk control, whereas, risk financing is the financing of the risk, be it self funding, third party funding, or other forms of insurance.

This chapter deals primarily with physical risk management. Risk financing is indicated in the model to show its importance in the risk management process, but will not be discussed in this manuscript.

Physical Risk Management

Physical risk management consists of identifying and assessing of risks and introducing the necessary controls to reduce these risks or minimize their consequences.

RISK ASSESSMENT

Risks cannot be managed until they have been properly assessed. The definition of risk assessment is: "The evaluation and quantification of the likelihood of undesired events, the likelihood of injury and damage, and an estimation of the results thereof."

The role of risk assessment is to provide the necessary information on which to make decisions regarding the cost-effective commitment of resources to prevent loss. Risk assessment also can be used to determine if appropriate action is acceptable where it is impractical to totally eliminate the hazards. Risk assessment will indicate where the greatest gains can be made with the least effort and which action should be given priority. This prioritization will bring about greater safety with the minimum level of effort.

Risk assessment is vital, as risks cannot be identified unless the hazards are first identified, after which the risk is then evaluated and only then can appropriate controls be put in place to minimize the risks.

THE THREE STEPS OF RISK ASSESSMENT

- Hazard identification
- Risk analysis
- Risk evaluation

Once this process has been completed, a total risk profile can be compiled.

Hazard Identification

A hazard can be defined as:

- A hazard is a situation that has potential for injury, damage to property, harm to the environment, or all three (high risk acts/conditions).
- A hazard is a situation or action that has the potential for loss.

The two main hazardous identification techniques are the *comparative* and the *fundamental* methods.

Comparative Hazard Identification

This uses checklists based on industry standards, or existing codes of practices, or by comparing the plant in question to similar plants. Depending on the plant design and nature of the process, checklists may provide sufficient comparison to identify the major hazards existing.

Fundamental Hazard Identification

This is based on considerations of deviation from original design. A hazard study is carried out to identify which events, or actions, may trigger off hazardous situations. Fundamental hazard identification techniques include such techniques as:

- Hazard and operability study (HAZOP)
- Failure mode and effect analysis (FMEA)
- Failure mode effects and criticality analysis (FMECA)

Hazard and Operability Study (HAZOP)

This study is a structured and systematic examination of a planned or existing process or operation in order to identify and evaluate problems that may represent risks to personnel or equipment, or prevent efficient operation. The HAZOP technique was initially developed to analyze chemical process systems, but has been extended to other types of systems and complex operations. A HAZOP is a qualitative technique based on guide words and is carried out by a multidisciplinary team (HAZOP team) during a set of meetings (Model 7.2).

A HAZOP study is a study carried out on planned or existing plants in an effort to identify operability and safety hazards.

Failure Mode and Effect Analysis (FMEA)

This is a process for hazard identification where all known failure modes of components or parts of a system are considered and the undesired outcomes are noted.

FMEA is a component-level review of a design to identify failure modes and their consequences (Model 7.3). For each failure mode, the study defines the effect of the failure, identifying any hazards to the operator, the environment, the public, or damage to the asset.

Item	Guideword	Possible Causes	Consequences	Action	Person Responsible
1	Overheating	Intake filter blocked	Device can overheat and trip out	Clean filter and move intake pipe	Maintenance foreman

MODEL 7.2   An example of a HAZOP study.

Item	Failure Mode and Cause	Effect	Corrective Action
The main generator	Loose connection, bearing failure or insulation breakdown	No electricity, no heat and no lights	Regular inspection of these items
Main transformer cooler	Leakage of coolant due to corrosion or damage	Overheating, transformer trip, no power transmission	Protect cooler from damage and replace coolant monthly

MODEL 7.3   Failure mode and effect analysis.

Each failure is subsequently analyzed to determine if it would be detected by the user or operator before causing an accident or damage, or whether mitigation controls are in place to prevent damage, escalation, etc. Where detection or mitigation is deemed inadequate, redesign requirements are recommended.

Failure Mode and Effect Criticality Analysis (FMECA)

This is a variant of FMEA and introduces consideration of criticality to rank the identified hazards based on risk. FMECA identifies the chances of failure as well as the magnitude of the consequences, which are ranked to identify the most critical aspects. The critical component is usually identified in a FMECA (Model 7.4).

Fault Tree Analysis (FTA)

This analysis endeavors to get a clear picture of the situation to provide for effective and permanent correction of a problem before it results in a loss. An FTA helps eliminate the risk before the accident happens. By discovering every factor that could contribute to a loss and by tracing its origin, the FTA is one of the most effective

Item	Function	Component (Critical)	Effect
The generator could fail	To provide electrical power	The brushes could wear out	No power would reach the exciter coils
Transformer failure	To step down the voltage	The over-current device fails	Transformer failure, loss of power to network

MODEL 7.4   Failure mode and effect criticality analysis.

tools for hazard identification. The diagram which is compiled describes the relationship between the events and the conditions that lead up to the event. The FTA recognizes that an event can occur if certain conditions prevail and that these conditions are caused by certain accidents.

Fault tree analysis is also very effective in determining the root cause of accidents and near miss incidents. It helps evaluate equipment failure and predict potential hazards. Fault tree analysis is a useful safety audit tool that uses very simple statements of fact and is very objective and realistic.

FTA is a top-down, deductive analytical method. In FTA, initiating primary events, such as component failures, human errors, and external events are traced through Boolean logic gates to an undesired top event, such as an aircraft crash or nuclear reactor core meltdown.

A fault tree analysis is not only an ideal method of hazard identification, but also of analyzing the risks contained in a process or a plant. The fault tree analysis encourages objective thinking and is effective in tracing possible causes of accidents, evaluating possible equipment failure, which leads to the prediction of an accident.

Event Tree Analysis

An analysis that starts from an undesired initiator, such as a loss of critical supply, component failure, etc., and follows possible further system events through to a series of final consequences. As each new event is considered, a new node on the tree is added with a split of probabilities of taking either branch. The probabilities of a range of "top events" arising from the initial event can then be seen.

The event tree analysis is similar to the fault tree analysis and enables management to prevent loss producing events as a result of risks within the system. The event tree analysis is a predictive method of analyzing risks.

So What If It Happens? (SWIFT)

The SWIFT is a systematic system of hazard analysis using brainstorming techniques. It considers the deviation, the hazard, and the cause and endeavors to determine consequence, if it happens.

Checklists

This is used for hazard identification by comparing existing plants and mines with experience based on a list of failure modes and hazardous situations based on past experience.

Hazard Survey

Another method of identifying hazards is a hazard survey, which is the application of existing loss prevention techniques to assess the hazards within an installation and to plan means of controlling them.

Hazard Indices

A hazard index is a checklist method of identifying hazards and ranking the degree of the hazard posed by the situation. This ranking helps prioritize resources in the form of manpower and capital expenditure.

Accident/Near Miss Incident Reports and Investigations

Another method of hazard identification is the examination of past accidents and near miss incidents to determine what hazards contributed to the injury or were involved in the event. Recurring accidents could indicate specific hazards in the process or the behavior of the people. After risk ranking, near miss incident investigation should receive the same attention as accident investigation and may prove to be more meaningful because it is predictive whereas accident investigation is reactive.

Thorough accident investigations will determine agencies and also isolate specific hazards as part of the overall risk assessment.

Critical Task Identification

Identifying critical tasks and analyzing them helps identify hazards in the process, method, or behavior. Obtaining information on near miss incidents occurring during the execution of critical or high risk tasks is essential. Once a task has been identified as a high risk or critical task, any near miss incident occurring should be viewed as important (as an accident) even though there was no loss. Potential for loss is the one criterion that initially helped identify the task as critical.

Safety Audit

A systematic and thorough safety audit (see Chapter 8) is also a method of identifying hazards and weaknesses in a safety management system. The safety auditing process quantifies the work being done to control accidental loss in the organization. This involves a physical tour and inspections of control systems that are intended to identify and eliminate hazards. Any shortcomings in current hazard identification processes will be identified by an audit. Therefore, audits are essential to an organization's hazard identification program.

Risk Analysis

Once the hazard identification process is complete and the hazards have been analyzed, a risk analysis follows.

A risk analysis can be defined as "the calculation and quantification of probabilities and consequence as a result of a risk."

Risk analysis is the scientific measurement of the degree of danger in an operation and is the product of the frequency and severity of undesired events. It sometimes views the probability, severity, and frequency of the event. It provides a predictive method of projecting the risk and analyzing the possible occurrence, the probability of occurrence, and estimating the consequences.

A risk analysis involves a *probability analysis, frequency analysis*, and an *impact analysis*. It helps reduce uncertainties as much as possible because it provides all available information on which to base risk reduction activities.

- A probability analysis asks: What could happen here?
- A frequency analysis asks: What is the exposure to this risk or how often could this occur or how many people, and how often are they exposed to the risk?

Probability	1 Low	2 Medium	3 Medium High	4 High
Severity	1 Low	2 Medium	3 Medium High	4 High
Frequency	1 Low	2 Medium	3 Medium High	4 High

MODEL 7.5 A simple risk analysis model.

- An impact analysis asks the question: How desirable or undesirable is the outcome of this risk?

Risk Ranking Approach

By taking into account the frequency and severity, a figure that is called the risk index can be derived. The formula for calculating a simple risk index is:

$$\text{Risk Index} = \text{Probability} \times \text{Frequency} \times \text{Severity}$$

Where probability is the chance of something happening, frequency is: How often could it happen? Severity is: How bad would it be?

Referring to Model 7.5, the probability, severity and frequency are ranked on a scale of 1 to 4 depending on the degree of exposure and size of consequence. The resultant figure is termed the risk index and gives some indication as to the ranking of the risk.

Purpose of Risk Analysis

Risk analysis is the measurement of the degree of danger involved in an operation or activity. The purpose of a risk analysis is to reduce the uncertainty of a potential accident situation by providing a framework to incorporate all risk eventualities.

Degree of Severity

To ensure that we analyze the loss exposure correctly, we should determine the size of the loss. There are two eventualities in the consequence equation: *the maximum possible loss* and the *maximum probable loss*.

1. The maximum possible loss is the absolute worst loss that could conceivably happen when one assumes a breakdown of all control systems.
2. The maximum probable loss is the worst loss expected to occur under normal circumstances if the existing control systems function reasonably well.

Another consideration could be the total loss effect of the event, which includes loss to people, property, and product, and also the costs of liabilities stemming from the event. In risk assessment the normal loss expectancy under normal conditions should be considered rather than the extremes.

Degree of Risk	Risk Score
Extremely high risk	100
High risk	70
Major risk	60
Minor risk	40
Acceptable or tolerable risk	30

MODEL 7.6 An example of a risk score.

Consequence Modeling

Taking into account the likely outcome of an event as well as the frequency of the event occurring, and considering the extent of losses, a consequence model can be derived from the risk analysis.

Risk Score

The most important objective of the risk analysis is to derive a risk score based on certain criteria that objectively can identify the greatest risks to help prioritize action. The risk score analysis (Model 7.6) takes into consideration the likelihood, the exposure, and the possible consequences and equates them to a risk score under the following categories:

Likelihood

The likelihood or probability of the event occurring as a result of the risk is measured on the following scale, as an example:

- May happen sometime
- Could happen
- Can happen
- Remotely possible
- Unlikely
- Impossible

Exposure

The exposure (frequency) or the number of times that the event may occur could be rated on a scale varying from annually to continuous. The rankings could be:

- Annually
- A few times per year
- Bimonthly
- Monthly
- Weekly
- Daily
- Anytime

Consequences

The consequences (outcome) are ranked on a scale from a catastrophe to minor interruption, for example. A catastrophe is where there are numerous fatalities and losses exceeding several million dollars damage. The rankings could be:

- Catastrophe
- Disaster
- Very serious
- Major event
- Noticeable
- Interruption

Risk Score Calculation

Ranking the risk being analyzed on the likelihood, exposure, and consequence scale determines the risk score, which could be from extreme (100) to acceptable (0–20). The risk analysis and risk score then enables us to prioritize action by using accepted risk control methods.

Risk Evaluation

Once the risks have been analyzed and risk scores determined, it is now possible to do a risk evaluation. A risk evaluation is a quantification of the risks at hand and an evaluation of the cost of risk reduction and benefits derived from reducing or eliminating risks. It is basically a cost-benefit analysis to determine which risks can be reduced, the benefit of risk reduction, and the cost of reduction.

One of the main objectives of risk evaluation is to enable management to take decisions on which risks should receive priority and where risk control efforts should be directed.

Risk Reduction

By taking the risk score, equating it with the percentage of risk reduction, and comparing the cost of risk reduction to the benefits, a thorough evaluation can be made of whether or not the risk reduction justifies the effort.

For example, it may cost a lot of money to improve the instrumentation on a control circuit. The cost of reducing this risk would not be justified because the possibility of an undesired event occurring due to the instrumentation is extremely slight and the consequence of the risk would be a minimal interruption only.

Acceptable or Tolerable Risk

In every walk of life there is a certain amount of risk. The same can be said for manufacturing and mining processes. It is virtually financially and physically impossible to eliminate all risks from all walks of life and, therefore, we accept a certain amount of risk as part of our day-to-day living. Safety has been defined by some as "acceptable risk."

As Low as Is Reasonably Practical (ALARP)

If the risks are kept as low as is reasonably possible (ALARP), it is accepted business practice. As soon as the risks extend beyond the ALARP region, then the consequences of those risks could be detrimental to the business and the people working there.

The only way to evaluate which risks should receive priority for rectification is by doing a thorough risk evaluation, weighing the risk score, percentage risk reduction, cost of correction, and the benefits that can be categorized as:

- Extremely beneficial: 100 percent
- Beneficial: 50 percent
- Not really justified: 25 percent

Those benefits that fall into the first two categories should receive priority. Consideration should be given to the nonjustifiable efforts as to whether they reduce the risk to the ALARP region or not.

This risk evaluation assists management in budgeting for risk reduction, helps prioritize expenditure and effort, and gives specific target areas for risk reduction efforts. Further risk reduction also is highlighted by the risk evaluation, and decisions can be made as to whether or not they are practical.

Decisions concerning the cost-effective commitment of resources to the risk control program are easily taken once the risk evaluation has been studied. Cost-effective risk reduction methods and time frames also can be compiled once the risk evaluation has been completed.

RISK CONTROL

If we can't live with the risk, what can we do about it? Risk control is the second step of physical risk management after the risk assessment.

The goals of risk control include:

- Preventing undesired events occurring.
- Avoiding risks occurring.
- Ensuring that preventative action is taken to reduce risks.
- Ensuring that contingency plans are in place should an undesired event occur.
- Transferring the risks wherever possible.

DEALING WITH RISK

To deal with the risks a risk management plan is put into operation. This risk management plan is often referred to as the nuts and bolts of risk control and includes the four ways to handle risk: treat, tolerate, transfer, or terminate.

Treat

In treating the risk, loss control, loss prevention, and safety and health management systems are implemented. This includes the safety management function of control, which will ensure that all potential hazards are identified, the work to be done is identified, standards are incorporated, and a system initiated.

Tolerate

Tolerate is an approach where the risks are financed and where the losses are paid for. The financing of these risks is normally done within the organization or sometimes with insurance agreements. Tolerating the risk, in some instances, is the only economical outlet in the presence of good risk control programs.

In 1978, a design flaw in Ford Pinto cars caused their gas tanks to explode in rear-end collisions that cost the lives of 27 people. A memo showed that Ford executives had weighed the cost of recall at $121 million against estimated lawsuits that may have cost only $50 million.

Under pressure from the Traffic Safety Administration and Ralph Nader, Ford eventually recalled 1.5 million vehicles in 1978 and scrapped the brand in 1980. They had initially decided to tolerate the risk.

Transfer

Transfer of the risk is when the losses are financed through insurance companies. Risks, which have a low frequency and high severity, are normally transferred. The first choice is normally self-insurance, but, depending on the organization, this could be insured by outside insurers.

Terminate

A risk control technique called *risk avoidance* is when the risk or process that causes the risk is terminated. The organization refuses to expose its employees and assets to the risk, so the risk is completely eliminated. Termination of the risk could mean stopping the process, removing the machine, or ceasing to use a hazardous substance, chemical, or process.

A risk management plan to treat, tolerate, transfer, or terminate the risks is now put into action. Treating the risks involves incorporating the safety management control function: ISSMECC (Model 7.7).

I	Identification of the risks
I	Identification of the work being done to reduce risks and combat losses
S	Setting of standards of measurement (what it must be like)
S	Setting of standards of accountability (who must do what)
M	Measurement against the standards
E	Evaluation of conformance's or deviations
C	Corrective action implemented
C	Commendation for good performance

MODEL 7.7 The management control function.

SAFETY MANAGEMENT CONTROL

A lot of effort goes into treating risks and this is where the safety management control function features. The safety control function or IISSMECC is shown in Model 7.7.

This safety control function is part of the risk management plan and is derived from the hazard identification, risk analysis, and risk evaluation, which are the three legs of risk assessment.

RISK RANKING OF NEAR MISS INCIDENTS

To determine which near miss incident should be investigated via the accident investigation system, each near miss incident should be risk ranked as to potential loss and frequency of recurrence. In Model 7.8 are a few examples from industry of reported near miss incidents that have been ranked on a simple "low (1), medium (2), medium-high (3), high (4) scale."

From these reports, it will be interesting to note that the area has been identified and although some reports are clearly high risk practices or high risk conditions, the

No.	Date	Area	Near Miss Incident	S	F	Potential Accident Type
1	2/1/2010	3740	Two welders received minor shocks when they touched a GMAW Welding Machine (No. M24). On inspection, a loose wire was discovered in the plug of an extension lead that was connected to the welding machine.	4	2	735
2	3/12/2010	3654	Employee was walking toward his job on the demarcated walkway inside big workshop. The forklift driver rode towards the big workshop. On the corner the driver nearly knocked Mr. Jones over, because he was unaware of the forklift approaching.	3	1	971
3	3/19/2010	3739	A miner was pulling out hoses to set up the jack leg and the hoses hung up making the miner mad. He pulled real hard and lost his balance and fell down.	1	3	740
4	3/30/2010	Offsite	I was pulling out onto the road (in the dark) and almost hit two bikers. They were descending the hill rapidly and had no lights or light clothing. I saw them at the last moment as shadows on the head lights of a car that had appeared over the top of the hill.	3	2	Road accident

MODEL 7.8   Near miss incidents risk ranking.

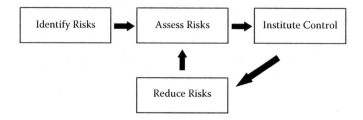

MODEL 7.9   The risk assessment process loop.

benefit of the system is that employees are reporting hazards and allocating a risk ranking to them.

CONCLUSION

Risk management consists of two components:

1. Physical risk management
2. Risk financing

Physical risk management consists of assessing the risks and then controlling them. Risk assessment consists of:

- Identifying the hazards
- Analyzing the risks
- Evaluating the risks

This then gives a total risk profile and priorities can be determined for control. Risk control includes the prevention of risks occurring, the avoidance of risk, and contingency plans should undesired events be triggered off by risks (Model 7.9).

8 Safety Auditing

Part of a memorandum the senior personnel manager (operations) of the London Underground wrote to the operating management meeting in August 1987 stated:

> A safe environment is not one in which there is an absence or a low number of serious injuries, but is the result of active participation by management and staff in identifying hazards and then doing something positive about them.
>
> In other words, the absence of accidents is a negative measure largely dependent on luck, while the identification then prompt elimination, or control of hazards, is a positive step and is essential to the discharge of our duties under current legislation.

In the light of the above statement, it is clear that proactive safety control is needed rather than reactive actions. Safety audits measure the management work being done to control losses and, therefore, are vital performance indicators in the safety system.

The near miss incident reporting and remedy (NEMIRR) system is a proactive, preventative safety initiative, and, as with all safety systems, should be audited on a regular basis. This will determine if the system is functioning correctly and that corrective actions are being implemented.

INTRODUCTION

Successfully managing the risk of accidents and injuries, for many organizations, means going beyond legal compliance. Safety management system audits are used by leading organizations worldwide to benchmark their safety management system against best practice processes. An audit objectively, and in detail, evaluates an organization's occupational health and safety management system, identifying areas of strength and weakness, and supports a structured continuous improvement approach going forward. A safety audit, by definition:

- Provides the means for a systematic analysis of each element of a loss control program to determine the extent and quality of the controls.
- Is a critical examination of all, or part, of a total safety operating system.
- Is a management tool that measures the overall operating effectiveness of a company's safety and health program.

REASONS FOR AUDITS

Most organizations only measure safety performance by the number of disabling, or lost-time, injuries that are experienced. This performance is normally in the form of a disabling injury incidence rate (DIIR) or disabling injury severity rate (DISR).

Regretfully, this measurement only measures the consequences of poor control. As they are largely dependent on luck, they do not accurately reflect the work being done to control losses.

The Health and Safety Executive (HSE) of Great Britain conducted a study of safety programs and published a report in 1976 entitled, *Success and Failure in Accident Prevention.* The study was conducted by the Accident Prevention Advisory Unit (APAU) of the HSE and the extensive study was summarized as follows:

> Any simple measurement of performance in terms of injury frequency rates or incidence rates is not seen as a reliable guide to the safety performance of an undertaking. The report finds there is no clear relation between such measurements and the work conditions, the injury potential, or the severity of injuries that have occurred. A need exists for more accurate measurements so that a better assessment can be made of efforts to control foreseeable risks. It is suggested that more meaningful information would be obtained from systematic inspection and auditing of physical safe guards, systems of work, rules and procedures, and training methods, than on data about injury experience alone.

The information on injury experience is reaction and not control. Audits measure the amount of control there is over the risks produced by the nature of the organization.

Tom Peters and Bob Waterman stated in their book, *In Search of Excellence:* "What gets measured, gets done." By putting a measurement on something, it is the same as getting it done. By quantifying safety, we focus attention to that area and once the information is made available people respond to it.

Measurement is a comparison with standards. Without adequate standards, there can be no measurement or evaluation of safety systems. Only the minority of all accidental events result in major injuries, some produce minor injuries, and about 4 percent cause property damage. The majority of accidental events do not result in injuries, and, therefore, control systems are essential to eliminate these near miss incidents, or warnings.

Past experience has shown that management always believes that their safety performance is much better than it actually is. Auditing the safety and health system is the only positive and progressive method of measuring safety performance and of indicating to management how well they are doing in safety in comparison with their standards and best practice.

BENEFITS OF AUDITS

A safety and health audit of the work being done to control loss measures the safety program and highlights its strengths and weaknesses. The audit compares work being done, and standards being maintained, with accepted safety and health standards.

A safety and health audit will help prioritize work to be done to further reduce the risks of the business. A systematic and thorough audit will indicate the development of a viable improvement plan, which can then be effectively communicated to all levels.

Audits also give recognition. Aspects of the safety program that are running well are scored and identified, thus giving recognition to good control. Audits also focus management's attention on the safety program elements and once its attention is focused on the safety elements, attention will be given to these elements.

The principle of *Safety Definition* (see Chapter 14, Investigating High Potential Near Miss Incidents) states that "decisions concerning the safety program can only be made if the basic causes of loss-producing events are clearly identified." An audit helps identify these basic causes. The audit helps direct the safety efforts. Executive leadership, commitment, and a desire to improve the work standards and reduce high risk behaviors lead to better safety, production, and higher quality standards. Only executives can set safety objectives that can be cascaded down to lower levels within the organization. These objectives are derived as a result of the audit.

HOW DOES AN AUDIT WORK?

Official auditors conduct a thorough physical examination of the entire work area. Relevant control documents are scrutinized, activities are monitored, and safety systems tested. The auditors critically examine each item of the safety program, test the system, interview employees, and quantify the work being done to control loss. The NEMIRR system would be audited for the following (Model 8.1):

- Is a formal system in place?
- Is there a written standard for near miss incident reporting and rectification?
- Have accountabilities for the NEMIRR system been defined?

4		INTERNAL ACCIDENT NEAR MISS REPORTING AND INVESTIGATION
	4.1	INTERNAL REPORT-DAMAGE AND NEAR MISS INCIDENTS
		Are all damage accidents recorded on an internal report form?
		Are details as required on the form accurately and completely filled in?
	4.2	INVESTIGATIONS BY DESIGNATED INVESTIGATOR
		Has an investigator been designated?
		Has the investigator been trained in the techniques of investigation?
		Are safety and health representatives involved in investigations?
	4.3	CAUSES IDENTIFIED
		Have the basic causes been identified and documented for every investigation?
		Have the immediate causes been identified?
		Do the basic causes conform to the accepted format?
	4.4	ACTION TAKEN
		Do investigation procedures incorporate positive recommendations for minimization or elimination of possible recurrence?
		Are the recommendations practical in terms of feasibility and cost consideration?
	4.5	FOLLOW-UP
		Is the recommended action taken within a reasonable time?
		Is a follow-up carried out after the action has been taken to determine if the action was effective and practical?

MODEL 8.1 An example of an audit protocol section for near miss incident reporting. A maximum score of five points is allocated for each question.

- Are forms or reporting methods available (report forms, booklets, computer stations, etc?)
- How many events were reported during the period?
- What is the feedback/notification system?
- Were these risk-ranked as to their severity?
- Has follow-up action been completed?
- What percentage of reports has been closed out?

WHO SHOULD CONDUCT AUDITS?

As many people as possible should be involved in the safety auditing process. This acts as a training process. Internally trained, accredited auditors could lead the audit teams. A more objective audit is obtained by having different auditors.

Health and safety representatives (HASREPS) should be used extensively for internal departmental audits and they also can participate in the biannual audit of the entire program. Official audits are conducted by external organizations that have auditors trained to conduct formal audits. This formal measurement also acts as a form of recognition for safety efforts. A level of performance is awarded or a star grading issued. The International Standards Organization (ISO) auditing protocols for environment and quality are examples of the international recognition given for achievement of standards.

Safety and health consultants can be used to conduct preliminary or baseline audits. Safety practitioners from similar organizations also could be invited to participate in the audit process. Management and employees should be well represented during audits and the safety coordinators and risk managers should be involved.

THE AUDIT PROGRAM

When embarking on a formal audit, the auditing program will follow a certain order (Model 8.2). The auditing process could be as follows:

- A pre-audit meeting is held and various element coordinators identified, physical areas chosen for the inspection, and selected employees are nominated for interviews. This pre-audit meeting also will act as a forum to ascertain who will be accompanying the auditors, the sizes and types of personal protective equipment required, and any other formalities that need be discussed.
- During the pre-audit meeting, the tour through the facility will be planned. The first day of the audit may entail examining the accident records, injury statistics, accident and near miss incident investigation, and recording systems. The safety organization of the program may be evaluated and policies, procedures, employment selection, and other criteria may be reviewed as well.

Action Plan Assignment Sheet for XYZ Engineering for the Recommendations as a Result of the Audit on 1 May 2010

What Is to Be Done?	Who Will Do It?	By When Must It Be Completed?
• Get all areas and all management to participate in the system. • Safety's roles, functions, and reporting to be revisited. • Address hazards listed in inspection report. • Continue producing Element Standards and their implementation. • Extend injury statistic tracking and reporting system to include non-injury events. • Calculate incidence rates for Divisions. • Investigation high potential near misses. • Agree upon and standardize vehicle prestart checks. • Identify high-risk tasks, conduct analyses, write procedures, and do critical task observations. • Set up a system for the reporting of property damage accidents. • Formalize incident recall sessions. • Introduce 6-monthly internal audits and allocate internal scores. • Continue with inspection regime and get Divisions to do inspections.		

MODEL 8.2   An example of an action plan derived from a safety audit.

- A safety orientation session should be held for the auditors before they enter the plant or mine to ensure that they follow the same standard induction process as new employees, vendors, visitors, etc.
- The next step in the auditing process is the physical inspection. The auditors may choose to spend between two and four days carrying out the physical inspection to ensure that the entire work site is covered. The duration of the inspection will depend on the size and nature of the facility been audited. Random sampling methods will more than likely be used.
- Once the physical inspection has been carried out, interviews are conducted with selected employees. These interviews will determine the depth of the program and also indicate whether or not safety systems and elements are in place and are working. The audit will include random sampling and actual counts of conformance and nonconformance to standards.
- Once all the documentation has been reviewed and systems are verified, a feedback session and presentation is held with all concerned parties. Here the auditors give feedback to the strengths and weaknesses of the program and allocate a percentage score to the entire safety system.

- After the audit, a formal audit report is produced and discussed by the auditors. This report will include proposed action plans to rectify the weaknesses found during the audit. Prioritization will take place and management will be left with an action plan listing what needs to be done and prioritizing the action to be taken.

HOW TO DO AN AUDIT INSPECTION

Most people make the mistake of walking through a work area too fast. Rather, one should stand at the entrance to the workplace and observe the people, machinery, environment, material, and work process. Notice what employees are doing in the area, the protective equipment being worn, and the environmental factors, such as the lighting, ventilation, noise, and temperatures. During the inspection always ask the questions, "Is this safe? Can this cause an accident? Is this according to standards?"

The auditors should try and cover the entire work area and always note and record deviations. All hazards must be recorded and brought to the attention of the employees in the area being inspected. While doing the inspection, be curious, look inside cupboards, look under workbenches, look on top of mezzanine floors, and look behind articles.

Always note the area in which the deviation was found or where compliance to standards was found. Maintain a positive attitude during the inspection, smile, be friendly, and compliment people wherever possible. Always record the deviations and where they occurred. If necessary react immediately and recommend that A-class hazards be rectified as soon as possible.

Keep an open mind and do not be biased during the inspection. It is important that all deviations are classified under the various elements during the inspection so that the evaluation can be conducted easily.

Always set an example by wearing the corrective personal protective equipment during the inspection and do not commit high risk acts, such as failing to observe a warning or mandatory sign.

INTERNATIONALLY ACCEPTED AUDIT-BASED SAFETY SYSTEMS

There are numerous, prominent international audit-based safety systems presently in use throughout the world. They include:

- The NOSA (National Occupational Safety Association) Five-Star Safety System
- The Safety Projects International (SPI), Five-Star Health and Safety Management System
- The British Safety Council's Five-Star Health and Safety Audit System
- The International Loss Control Institute's International Safety Rating System (ISRS)
- The International Standards Organization (ISO)
- The British Standards-Based Occupational Health and Safety Assessment Series (OHSAS)

OCCUPATIONAL HEALTH AND SAFETY ASSESSMENT SERIES (OHSAS)

The OHSAS-type system was largely developed from the safety rating systems mentioned. OHSAS 18001 is compatible with ISO 14000: 1997 (Environmental) and ISO 9001: 2000 (Quality) management system standards.

NATIONAL OCCUPATIONAL SAFETY ASSOCIATION (NOSA)
AND SAFETY PROJECTS INTERNATIONAL (SPI)

The NOSA and SPI Five-Star Systems have 73/78 critical areas under five main sections:

- Premises and Housekeeping
- Mechanical, Electrical, and Personal Safeguarding
- Fire Protection and Prevention
- Accident Recording and Investigation
- Health and Safety Organization

The total points allocated vary with the audit systems, but are all equated to a percentage to indicate what star grading or score is awarded.

Under the NOSA rating system, the performance evaluated is based on 12 months past experience and the star grading is only valid for 12 months. To achieve a star rating, a percentage effort must be obtained and the DIIR must be below a certain figure. Different systems have slightly different scores (Model 8.3).

THE INTERNATIONAL SAFETY RATING SYSTEM (ISRS)

A widely used system, which sets standards of measurement, is the International Loss Control Institute's (ILCI) International Safety Rating System (ISRS). (ILCI was incorporated into Det Norske Veritas (DNV) in 1995.) This system also has set standards of conformance as well as standards of accountability for safety work. It incorporates physical inspection guidelines and precise instructions to accredited auditors on how to allocate scores. In the ISRS there are 20 main element headings, each one composed of a number of elements. The total score is 13,000 points, though updated systems may have different scores.

No. of Stars	Grading Percentage	DIIR
5 — Excellent	91%	1
4 — Very Good	75%	2
3 — Good	61%	3
2 — Average	51%	4
1 — Fair	40%	5

MODEL 8.3   Star-Grading Criteria: The audit percentage and injury rate needed for a star grading under the NOSA system. (From NOSA Grading Criteria, 1991.)

These pioneering auditing systems are both systematic and thorough in their approach and accurately quantify the work being done to control loss while giving recognition for the safety effort. There are other similar systems in use, which operate in a similar way.

OCCUPATIONAL HEALTH AND SAFETY ASSESSMENT SERIES SYSTEM (BSI–OHSAS 18001)

A health and safety management system is a formalized approach to health and safety management through use of a framework that aids the identification and control of safety and health risks. Through routine monitoring, an organization checks compliance against its own documented health and safety management system (SMS) as well as legislative and regulatory compliance. A well-designed and safely operated management system reduces accident potential and improves the overall management processes of an organization.

British Standards based, (BSI–OHSAS) 18001 is the occupational health and safety assessment series specification. Implementation demonstrates an organization's commitment to protection of health and safety in the workplace. It comprises two parts, 18001 and 18002, and embraces a number of other publications.

Key principles include:

- Leadership and management: clear commitment
- Setting objectives: continual improvement
- Planning: hazard identification, risk assessment, and risk control
- Competence: training and awareness
- Consultation and communication: all stakeholders
- Structure and responsibility: clear lines and definitions
- SMS: audit and review to monitor effectiveness

BSI–OHSAS 18001 was specifically developed with requirements of the ISO 9001 Quality Management System and ISO 14001 Environmental Management System in mind, allowing for ease of integration of management systems.

OHSAS 18001 was created via a concerted effort from a number of the world's leading national standards bodies, certification bodies, and specialist consultancies. A main driver for this was to try to remove confusion in the workplace from the proliferation of certifiable occupation health and safety specifications.

CONCLUSION

Most safety programs measure their consequence, which is postcontact control and reactive safety. Audits evaluate every aspect of work being done to prevent accidental loss and are ideal means of measuring precontact safety control efforts. A safety audit is the best means of identifying whether a safety system and its near miss incident reporting system is functioning according to standards. Management is more inclined to pay attention to anything that is quantified and the same applies to safety.

9 Near Miss Incident and Accident Recall

INTRODUCTION

It has often been said that past experience usually predicts future trends. The same can be said for safety. Accidents that have occurred in the past tend to recur and the same basic causes also are present. Numerous near miss incidents often predict accidents.

Recalling the past to improve the future is of vital importance in a safety management system and the reason that most programs fail is that the underlying near miss incidents are never reported, investigated, or eradicated.

Numerous safety programs are still injury prevention programs and only concentrate on the serious or disabling injury and the minor injuries as indicated in the accident ratio models.

The accident ratio (first discussed by H. W. Heinrich in the 1930s [1931]) has a clear message:

> Accidents and not injuries should be the point of attack. Analysis proves that for every mishap resulting in an injury there are many other similar accidents that cause no injuries whatsoever (p. 24).

For every 641 undesired occurrences, 1 will result in serious injury, 10 will result in minor injury, 30 will end up causing property and equipment damage, and some 600 will have no visible outcome or consequence (based on the Frank E. Bird [1992] ratio).

The basic philosophy of accident prevention is that if you look after the near miss incidents, the injury and damage-causing accidents will look after themselves.

Near miss incident recall and accident recall are ideal methods of recalling various accidents and near miss incidents in an effort to remind people of what went wrong in the past and also to highlight potential near miss incidents that may have not been reported via the formal system. It is a method to learn from past mistakes.

REPORTING

Once an integrated safety system has been implemented, the reporting of injuries and property damage accidents is normally good. The reason for this reporting is that employees and management can understand and see the resultant injury or property damage. They understand the loss attached to these accidents, report them, investigate them, and take action to prevent their recurrence. This is also a worldwide legal requirement prescribed by safety laws.

Near miss incidents with no visible outcome or resultant injury or damage are often not regarded as important. They are not seen or maybe not recognized by the average employee.

Accident recall is the ideal way of reminding people. By recalling past injury or damage-causing accidents and bringing about an awareness of their causes, we ensure that steps are in place so that a recurrence does not happen.

Near miss incident recall, on the other hand, is a method of recalling near miss incidents that did not result in any injury, visible damage, or production loss, but which may have if circumstances were different.

Definitions of an accident and near miss incident are:

- *An accident* is an undesired event, which results in harm to people and/or property damage and/or process interruption.
- A *near miss incident* is an undesired event, which under slightly different circumstances may have caused harm to people and/or property damage and/or process loss.

The difference between the accident and the near miss incident is purely a matter of chance as the outcome of a near miss incident cannot be determined and is very difficult to predict.

A MATTER OF LUCK

SCENARIO 1

A worker is walking under a scaffold on a building site and a brick is knocked from the scaffold above. The brick falls 20 meters (70 feet), but lands undamaged a few centimeters from the worker.

This is a near miss incident as there was no visible loss, no injury to the worker, and it did not interrupt the business process.

SCENARIO 2

A worker walking under scaffolding is struck on the hand by a brick, which fell 20 meters (70 feet) from the working area on the scaffold above. The injury is a minor graze to the back of the hand and requires light first aid treatment.

This is an accident as the end result of the event was physical injury, but once again no loss occurred as the brick was undamaged.

SCENARIO 3

In this case, the brick also fell 20 meters (70 feet), but due to chance, the worker happened to be right in its path. The brick fell onto his neck close to his head, knocked him unconscious, and caused a severe concussion that hospitalized him for a month.

This is an accident as it resulted in physical injury to the person.

SCENARIO 4

The same brick falls, narrowly missing the worker, but crashes down onto a portable generator being used to power a drilling machine. The machine's fuel tank is ruptured by the damage and has to be shut down and repaired. This is a property damage and business interruption accident.

CONCLUSION

The brick falling was an undesired event. When it missed the worker, it is classified as a near miss incident as there was a flow of energy, but it caused no loss. It was caused by unsafe stacking of bricks on the scaffolding. If the brick had fallen three times with no damage or no injury, we would have had three similar near miss incident events.

The fact that the brick missed the employee in Scenario 1 defines it as a near miss incident as it had no visible effects, such as damage or injury to people or process disruption.

The undesired event of the brick falling in Scenario 2 resulted in minor injury and, therefore, is classified as an accident as it did cause harm to a person.

The undesired event in Scenario 3 is also an accident except that this time the accident caused severe injury in comparison to the minor injury in Scenario 2.

Scenario 4 is clearly a damage accident as there was damage to the generator.

This indicates that most near miss incidents have the potential under slightly different circumstances to cause injury, property damage, business interruption, or a combination of all of these losses.

The difference between Scenarios 1, 2, 3, and 4 is a matter of chance, or luck. The worker was lucky in Scenario 1 and 4, not so lucky in Scenario 2 and had real bad luck in No. 3 as the slightly different circumstances put him in the path of the falling brick.

The luck factors could be described as positioning, timing, or being in the wrong place at the right time.

RECALLING THE NEAR MISS INCIDENT

The near miss incident, the undesired event of the brick falling but with no damage or consequence, would form one of the plenty of near miss incidents, which eventually lead to property damage and injury-producing accidents. Near misses are often the foundation of major injuries. They are the same as accidents except for the missing phase of the exchange of energy above the threshold limit of the body or structure.

Near miss incident recall is a technique used to recall these near miss incidents that had potential to cause loss, but were never identified nor reported.

The falling brick was caused by a high risk condition. Most near miss incidents that will be recalled are either high risk conditions or high risk acts, or a combination of both. Although these are not pure near miss incidents as per definition, their reporting is important. They cause the undesired event, but, because of the circumstances, do not result in any visible loss. Once a high risk act or condition results in a flow of energy, we have a near miss incident. A falling brick creates a flow of energy.

BENEFITS OF NEAR MISS INCIDENT RECALL

The importance of near miss incident recall cannot be emphasized enough. Accidents that cause injury, damage, and production losses are recognized as being important and are reported and investigated. Many near miss incidents are ignored and not reported or acted on.

Near miss incident recall reveals the events that could have injured someone or damaged something. Near miss incident recall is the ideal way of getting near miss incidents reported, so they can be investigated and positive steps can be taken to eliminate the root causes.

Near miss incident recall also can offer a guide as to what training is needed concerning the high risk acts and conditions that are being reported. It can help to indicate certain trends as well.

BENEFITS OF ACCIDENT RECALL

Accident recall revisits previous work injuries or occupational diseases or damage accidents, which have already been investigated and analyzed. It is also a proactive safety measure, based on postcontact events, and it acts as a reminder of accidents that happened in the past. It also offers a guide to ensure that follow up is still in force as a result of previous accident investigation recommendations.

Accident recall gets active participation from and shares valuable experience with the group. It helps to motivate as it indicates that past accidents are not merely forgotten, and rated unimportant, but are recalled frequently and constant vigilance is thus maintained.

Accident recall is ideal for induction training and also for making the new worker aware of the type of accidents that have occurred in the past and that may occur in the future.

PRECONTACT AND POSTCONTACT ACTIVITIES

Near miss incident recall is a precontact activity. It allows the recall of near miss incidents that have not yet caused harm or damage and endeavors to identify the root causes and rectify them before any contact takes place.

Accident recall is a postcontact safety activity, as a loss must have occurred before the accident could be recalled to remind workers of its details.

DISCIPLINE

Employees will be hesitant to recall near miss incidents if the recall ends up in disciplinary measures. A near miss incident recall session should be treated as confidential and the information disclosed also should be handled with discretion. No disciplinary measures should take place. Finger-pointing and punishing people for recalling near miss incidents will stop employees participating in the near miss incident recall sessions. Where possible, the near miss incident recall should be a fact-finding session and not a blame fixing or witch hunt activity.

METHODS OF RECALL

There are various methods of conducting both near miss incident and accident recall and the two main methods are the *formal* and *informal* recall sessions.

FORMAL RECALL

Formal recall can be added as an agenda item at safety and health committee meetings or be part of the work carried out by a small group. At the end of training sessions, a five-minute recall session could be held and any near miss incidents that were noticed could be recalled by the employees attending the training session.

INFORMAL RECALL

Informal recall sessions can be held on a person-to-person basis during the normal workday. A near miss incident recall form also could be available and anybody could then fill in the form and report a near miss incident. This report could remain anonymous and the person reporting the near miss incident need not name the person or persons involved.

The form shown in Model 9.1 is a simple yet effective reporting form for near miss incidents, high risk situations, or behavior. It incorporates a simple form of risk-ranking and allows for anonymity of reporting.

Employee Report of a

SAFETY OBSERVATION

Near Miss Incident	High Risk Condition	High Risk Act	Safe Work

Employee Name (Optional): ....

Potential Severity	Major (3)	Severe (2)	Minor (1)
Recurrence Rate	Frequent (3)	Occasional (2)	Rare (1)

Description of
event:_____

_____

Corrective Actions:_____

_____

_____

Follow-up action:

Who Promised	To Whom	Action Promised	By when?	Date Completed	Issue resolved

MODEL 9.1   Example of a near miss incident reporting form.

MAJOR LOSS BRIEFING

A major loss briefing is an accident recall session that takes place after a major accidental loss. This major loss could be a fatal or permanent injury, serious property damage, or a combination of both.

A description of the accident is presented at the meeting or gathering and copies distributed to everybody in the plant to keep them informed. This description also is pinned up on the various notice boards so that complete exposure is achieved and awareness is created among the employees.

This major loss briefing is an ideal way to make the workforce aware of the causes of the accidents, and also the remedial steps that have been taken to prevent a recurrence.

SAFETY STAND DOWN

After a fatal accident, many organizations down tools and have a safety stand down. This is when work stops; employees are grouped in various halls or lecture rooms and are addressed by management and union leaders. The discussion is almost a last ditch stand in an effort to stop high risk situations that have already had dire consequences. It is used as a major "wake up call" to all that what happened is unacceptable and that the situation must improve from now on.

The stand down is a company-wide accident recall session. The details of the event, normally a fatal accident, are recalled, analyzed, and a line is drawn in the sand with management demanding an improvement in safety cultures, habits, and performance from that point on. Regrettably, the stand downs are normally held after the event rather than before the event.

SAFETY STAND DOWN BASED ON NEAR MISS INCIDENTS

Below is a report of a safety stand down that was held involving some 300 mine employees.

SAFETY STAND DOWN DAY AT MINE OPERATIONS, 12, 14, AND 15 MAY

Under the leadership of the Mine Manager, the entire development section of Mine Operations, consisting of some 300 employees participated in an 8-hour safety stand down. The stand down covered all employees within the division and took place on the A shift on Tuesday, 12 May, B shift on Thursday, 14 May, and C shift (Graveyard) on Friday, 15 May.

The very intense program was initiated as a result of serious injuries being experienced in the development section and also as a result of the types of near misses being reported from the division.

The stand down commenced with a classroom-based welcome session where the Vice President and Group General Manager of Operations welcomed all and expressed management's support for the entire safety effort, and their personal safety at work. The Mine Manager then explained the purpose of the event and set the competitive scene with an announcement of a competition between the teams, the winning team over all three shifts being awarded a free pair of safety boots. The Joint Union Management Safety Coordinator made a presentation followed by the Manager of Safety and Security.

Three main venues were set up for the activities; classroom, surface and underground. An hour was spent in the classroom where corrective actions, hazard recognition and ranking, 5-second risk assessment, unsafe conditions and acts, and planned job observations were taught. The 4 surface locations taught near miss reporting, written safe work procedures, fall protection and LHD safety. The 4 underground locations consisted of an on-site explanation of underground geology, ground support and hazard recognition.

Each team was identified by a colored badge and were ranked, or scored, at each location. The event was finalized by a classroom get together where the teams had to list the 5 major hazards noted on their walk into the mine. This test was the final score to be added to their score cards. A feedback form which called for comments and suggestions was then handed out. A personal safety commitment goal card was also issued for each team member to set their own safety activity goals.

A representative of the State Mine Inspector's office then addressed the group on safety and the legal aspects after which the Mine Manager thanked the presenters and the participants, announced the winning team, and closed the proceedings.

ACCIDENT RECALL AIDS

One of the ideal methods of recalling accidents is to take the trainees on a tour of the organization's safety museum. The safety museum normally consists of unsafe hand tools, poor electrical equipment, dangerous hand tools, poor personal protective equipment, etc., and also contains photographs and descriptions of previous accidents. By touring the museum and recalling the stories behind the exhibits, an ideal accident recall session takes place.

A book containing photographs and descriptions of previous accidents also could be used to recall past accidents. This book can be used during induction training and occasionally used as well during regular safety committee meetings, group discussions, or small group activities.

SAFETY DVDS

Many organizations show specific safety DVDs to remind employees of past accidents in an effort to raise awareness. The showing of a film, and relating it to, either a near miss incident, or accident recall, is a good idea. Safety DVDs often have high risk acts or high risk conditions portrayed and these scenes could act as reminders and prompts for the accident or near miss incident recall session.

A safety DVD should always be introduced and a brief summary of the film given before its viewing. After the viewing, it is essential to summarize the contents of the film and to ask questions of the audience to determine the level of understanding of the message. At this time, near miss incident recall can be used. The various near miss incidents recalled or accidents recalled from the past that get discussed should be noted.

NEWSLETTERS

The safety newsletter also is a good medium for conveying the accident recall message. Regular articles should feature past accidents, their causes, and what was done to prevent a similar accident recurring. Newsletters can be used to encourage near miss incident recall by explaining its purpose and encouraging the readers to fill in an anonymous near miss incident recall report.

Emphasis should at all times be placed on the fact that near miss incident recall is fact-finding and not fault-finding and that no disciplinary measure whatsoever will follow the report of a near miss incident.

ACTING UPON NEAR MISS INCIDENT RECALL

The first step of the near miss incident recall process is to get the near miss incidents recalled, discussed, and recorded. Once the near miss incidents have been documented, follow-up action is vital to ensure that the hazard reported was assessed and the wheels set in motion for corrective action to be taken.

THANKS

Appreciation for the recalling of the near miss incident should be shown to the person who has contributed by recalling the near miss incident. By acting on an anonymously reported near miss incident indicates the sincerity of management and the efficiency of the near miss incident recall system.

RISK ASSESSMENT

A risk assessment of near miss incidents recalled should be done to rank the potential loss, the severity of the loss, and the frequency at which the loss may occur. The ranking of the near miss incident also will determine whether the rectification should receive an A priority, a B priority, or a C priority.

A. Indicates imminent danger to life and limb
B. Indicates the possibility of severe injury and costly damage
C. Indicates the possibility of minor injury and less costly damage

Some hazards or high risk practices may take longer to rectify than others, so some form of ranking system is advisable in order to prioritize. One organization uses a simple traffic light system to prioritize near miss incident action. Red indicated immediate action, Orange initiates urgent action, and Green means that the actions can be scheduled for rectification.

REMEDIES

There could be numerous remedies for the near miss incidents recalled and the particular remedy applied should treat the root causes of the near miss incident rather than the immediate causes.

IMPLEMENTATION

Irrespective of what remedial steps are to be implemented and what action is to be taken to rectify the root causes, some action must take place and people must be instructed to take that action within a certain time period.

The first part of near miss incident recall is getting the near miss incidents reported. The second part is the follow-up action that is vital to fulfill the main function of near miss incident recall, which is taking action on the near miss incidents before waiting for them to cause injury or damage.

Follow Up

After remedial action has been instituted, there should be follow up action to ensure that the root causes of the recalled near miss incident has been properly addressed and eliminated. Some remedies, such as training or modifying the process, may take longer, in which case, an ongoing follow up procedure should be initiated.

Near Miss Incident Recall Examples

- Apparently, a can of insect spray is not sufficient when tackling a wasp's nest. Many people like to do the job thoroughly, efficiently, and quickly, but not so safely. Merely light a flame a short distance in front of the spraying insecticide and you have a most effective blowtorch. With wasps, you can never take chances.
- A contractor carrying out construction work on a welding site was discovered to be welding with no eye protection whatsoever. The electrode holder was a pair of pliers that had been connected to the main cable and the eye protection that he was using was a black and white photograph negative.
- During a near miss incident recall session, a candidate recalled how the washing machine kept on tripping the earth-leakage device, GFCI (ground fault circuit interrupter). He told us that only his wife used the machine. To fix the problem, he disconnected the ground wire from the plug top of the washing machine. The GFCI stayed in the ON position and the washing machine worked. When I asked him how long ago he had disconnected the ground (safety) wire, he stated, "Oh, about four years ago."
- Oh! So, you've never heard of acetylene bombs? We have fun with them in the workshop every week. You simply adjust your oxygen and acetylene-welding torch until you get a nice flame, extinguish it, and pump the flammable mixture into any plastic bag you can find. Once the plastic bag is full, you insert a piece of string or similar fuse, tie off the neck and you have an acetylene bomb. Sometimes they go off with such a loud bang that they shake the dust off of the rafters. They also give the guys in the workshop one heck of a fright.
- A man who had recently attended a safety training course had to carry out work on his steeply pitched tiled roof. His training taught him that he needed some form of lifeline while he was on the roof, so he obtained a long length of rope, tied a nonslip bowline knot around his waist and sought a suitable anchor. The only readily available hitching point was the fender of his wife's car standing in the driveway. He tied the end of the rope to the fender firmly and proceeded to climb upon the roof. Taking up the tension

he felt quite secure and he started working. His wife, unaware, that her car was being used as a lifeline anchor, jumped into the car and sped off for her weekly hair appointment. The safety-conscious roof repairer was dragged halfway to the hairdresser before his wife realized that something was wrong with the pulling power of her car.

- A father fitted new carpets in his car. A few days later, he discovered to his astonishment that the newly fitted carpet was jamming the accelerator pedal causing the car to maintain a high speed. The first time it occurred, he vowed to refit the carpet so that it did not foul the accelerator pedal. While driving, he merely bent down, pulled the carpet back and this freed the accelerator. His son borrowed the car for the day having to drive some 50 miles for a job interview. On the way, the accelerator pedal jammed. The son bent down to grab the carpet to free the accelerator and momentarily took his eyes off the road. When he looked up he found that he had drifted off of the road and he jerked the steering wheel to bring the car back onto the road This caused the car to roll three times. Fortunately he was not killed, but only suffered broken ribs and a severe bruising.

- After heavy rains, a house owner was confronted with a tripping earth-leak-age device (GFCI) in the main electrical distribution board. He proceeded to open the distribution board and checked the circuits. He discovered that the circuit breaker feeding the swimming pool tripped the GFCI. He switched all the switches off, gathered his tools, and proceeded to the swimming pool where he noticed that the pump had got wet during the rain. While opening the electrical terminal cover of the pump, he suddenly found him-self lying on his back some 10 feet away from the pump housing. His wife had returned home and on noticing that the lights were out had switched on the electricity.

- While driving past the industrial area, we spotted an employee riding on the forks of a moving forklift truck travelling down the road. He was supporting the fridge being moved by the forklift truck.

- An employee recalls his accident:

I work in the Galvanizing Department at a large galvanizing factory in Nigel. I became a safety achiever for the sole reason that I was wearing my personal protective clothing when the accident happened.

In the Galvanizing Department, work is done in dangerous conditions and even in these conditions a very low accident rate is experienced. We galvanize different steel structures in hot zinc at the very high temperature of 460°C.

On this particular day we were busy with a structure in a V shape. It was giving some trouble in the sense that it was picking up dross (deposit at bottom of bath) and the beam needed to be galvanized several times to get rid of the dross.

Because of constantly dipping the structure in the zinc, it started to cool down and the cold structure in the hot zinc caused the zinc to splash the instant the beam was dipped into it.

I wasn't fast enough to get away, but I could turn around and was only hit on the right side of my back, on my arm, on my neck, and some in my face. I was rushed to hospital and was held overnight for observation. I went home the next day.

One thing I learnt through this accident is that you should wear your PPE (Personal Protective Equipment) at all times because you don't know when an accident might just happen to you.

NEAR MISS INCIDENT RECALL CASE STUDIES

Case Study 1

To do a safety presentation, I arrived at the lecture venue early one morning. Upon entering the room, I noticed that the lights were switched off and that the area was in darkness. Looking up, I immediately saw why. An electrician's helper was standing on the very top of a self-standing stepladder and was removing the fluorescent tubes from the light fitting. I assumed that he was busy with the replacement of the tubes. I returned to my car to collect another load of books and course handouts.

As I re-entered the training room, carrying a box full of books, the training instructor walked in behind me and said, "Why don't you switch the lights on?" With that he threw the light switch.

The entire training room suddenly lit up with all the lights having been switched on to reveal a startled and terrified electrician's helper with his hands full of electrical wires coming from the workings of the fluorescent lamp and from the main power supply. I'll never forget the look on his face. His eyes were wide open and he was frozen almost as if he had been turned momentarily into a statue.

In that split second, it was obvious that he had received the fright of his life as he knew he was working with current-carrying conductors and he had failed to isolate the circuit breaker or lock it out or tag it.

Case Study 2

This is a true story that can be used as an accident recall to emphasize the importance of home safety.

Born Out of Tragedy

Sue threw household methylated spirits (denatured alcohol) on a fire she thought was dead and it became a screaming fireball within seconds. Now John has to raise the baby and their other children alone. The grief-stricken father hugs little Franklin to his chest and says: "He's my last link with Sue ..."

> It was a rare sunny day and Susan Guest decided to have a barbecue in the backyard of her house in Bristol, England. "Let's surprise Daddy," she suggested to her two young children. "Let's make a fire, barbecue some meat, and by the time Daddy comes home it'll all be ready."
>
> In due course, the flames died down and the coals became covered in a layer of grey ash. Susan thought the fire had gone out and decided that a shot of spirits would get it going again. So, she poured a generous quantity of spirits over the coals. But, what she didn't know was that beneath the innocent-looking layer of ash those coals were white hot. The moment the spirits hit the coals, it looked as if an incendiary bomb had exploded. Before the eyes of her children, 4-year-old Karina and 15-month-old Calvin, Susan was enveloped in a pillar of flame.

The children screamed as she ran to their paddling pool, plunged in to quench the flames, and then ran into the house to phone for an ambulance.

Susan could see her reflection in the big mirror beside the telephone and knew it was serious, about 40 percent of her body was burned and blackened. And, the thought that plagued her most urgently was: "What's going to happen to the baby?" Susan was seven months pregnant.

Her neighbor had heard her screams and jumped over the wall. He found Susan at the telephone, led her gently upstairs and made her lie down in a bath filled with cold water until the ambulance came. When John Guest saw his wife in the intensive care unit, he hardly recognized her. She was attached to a respirator and covered in bandages.

"She was swollen to three times her normal size," he says. "It was terrible. It was difficult to believe that an ordinary barbecue could so seriously maim someone you love."

Two weeks later doctors decided Susan's condition had improved to the point where she could be disconnected from the respirator. Hours later she went into labor and had an emergency caesarean.

Susan's son was seven weeks premature, weighting only 3 pounds. Without Susan even having had a chance to see him, he was rushed to another hospital. The medical team just shook their heads and hoped for the best. It didn't look good. ...

John visited both hospitals as many as three times a day to check on Susan and baby Franklin. He couldn't hold his wife or kiss her because her wounds were too severe. Repeated skin grafts failed and she grew steadily weaker.

"The grafted skin kept bleeding," says John. "And then she developed a lung infection. Eventually there was little doctors could do for her and I just sat and held her hand until she died, never having seen or held her son."

He remembers Susan as a lovely wife, a bundle of energy, and he's still far too upset to pack her things away. "I know I have to sort through her stuff, but I keep postponing it," says John. "I think I'll keep her wedding dress and then let Karina have whatever she wants to remind her of her mother."

John didn't allow his older children to attend Susan's funeral, but he'll ensure they never forget her. He takes Karina and Calvin to see her grave every Sunday. "We call it Mummy's garden and leave flowers there every time we go."

To John, baby Franklin is a more tangible memory of Susan. He'd thrived despite his low birth mass and at nine weeks had been allowed to go home.

"He's my last link with Sue," says John. "Every time I look at him or the other children I see her in them. A part of me died when Sue died; I wish I'd died instead."

"My children need a mother. I could never be a mother to them; they worshipped Sue. Karina looks just like her." John recalls bursting into tears as he told Karina her mummy wouldn't be coming back. "I told her she'd gone to live with Jesus. I was in tears and she cried too."

"Calvin can't talk, but I know he longs for her desperately. He used to cry pitifully every time she so much as walked out of the door. He followed her everywhere."

John, who used to work in the building trade, now has to play the dual role of father and mother to his children. He can no longer work and is battling to keep his head above water. His day begins when Calvin or Franklin wakes up. He dresses the children and makes their breakfast, and he takes Karina to school. During the day he tries to iron, does the housework, shopping, cooking, and gives Calvin and Franklin as much attention as he can.

At 3:15 p.m., he fetches Karina from school and plays with her until dinner time. Before Franklin came home from hospital, he'd watch a little TV after dinner, but there's no time for that any more. For a man accustomed to working outdoors all day and coming home only at nightfall, it's taken a great deal of adjustment. He does have some help, however. Welfare officials have taught him to change diapers and make a bottle. Susan's two sisters, her parents, and John's parents also visit often.

"I'm doing my best and Franklin is going to keep me busy," says John. "I just hope I can bring him up as well as Sue would have. If only she'd waited for me to come home and do the barbecue. ..." (*YOU* magazine, 1994)

CONCLUSION

Near miss incident recall is a useful tool to get employees to share information about near miss incidents that they may recall from the workplace, home, or during recreation time. We tend to learn from our own mistakes and incident recall helps us learn from others mistakes as well.

If the incident recalled is not a pure near miss, but a high risk act or condition or even a loss-producing accident, it is still a reminder of what can happen, perhaps with drastic results next time. Employees working safely or seen doing an act of safety also should be acknowledged during recall sessions.

10 How to Motivate for Safety

INTRODUCTION

One of the obstacles most frequently encountered when introducing or managing a near miss incident reporting system is a lack of motivation from management and the workforce to embrace the new kid on the block.

This lack of motivation is sometimes experienced by the top management team, the workers, or certain divisions and departments within the organization. Introducing something new is always viewed as extra work and, in safety, it's a case of: "Oh, that won't work anyway."

MOTIVATION

Motivation is a two-way street and the question to be asked is: "Have we really tried?" In implementing a near miss incident reporting system, it is vital that the organization buy into the system and that line management support it and employees participate in the process. The thought starters listed below may give some assistance in helping to motivate others toward participating in the near miss incident reporting system.

CREATE THE RIGHT ENVIRONMENT

It has often been said that you cannot motivate other people. Motivation comes from within and others can only create an environment that is conducive to self-motivation. The three main factors that assist in creating this environment in which people can become motivated are: *involvement, inspiration,* and *impelling* people to greater heights.

INVOLVEMENT

To involve people, they must get ownership of a portion of the safety program. The near miss incident reporting and remedy system (NEMIRR) is ideal in this respect as each individual can participate freely in the reporting of near miss incidents without fear of reprisal. To motivate employees, they must participate in the decisions concerning the safety and health system and, where possible, take ownership of a part of the safety program. The near miss incident reporting mechanism gives all employees a role to play in the safety process.

For example, by appointing a carpenter responsible for the purchase, inspection, and maintenance of all the wooden ladders in a plant will give him ownership of a part of the safety and health system and thus get his involvement. Managers, not the safety person, should chair their departmental safety meetings. This will get managers involved and not regard safety as something that belongs with the safety department.

INSPIRE

Always encourage people by complimenting them and wishing them well for actions toward safety. When we inspire people, we should influence them by our own example. We should try and create a sense of excitement around the safety system. Safety competitions are specifically for the creation of the excitement of competition. The NEMIRR system introduces a competition in the form of peer pressure as employees don't want to be left out of the action. Once reports start coming in, they will want to compete and also report.

Submitting a near miss incident report is almost the same as sending an e-mail directly to the executive manager of the organization.

IMPEL

Where we can impel people toward safety we will then urge them forward and help them use their creative power to contribute to safety. Motivation is a two-way street and if we are not creating the environment that inspires employees to greater involvement in safety processes, we should not always point a finger at them and say they are not motivated. Our own personal test is: do we always encourage, involve, inspire, and impel employees? Do we do enough to recognize "safe work" reports and other near miss incident reports?

THE HAWTHORNE EFFECT

Between 1927 and 1932, at the Hawthorne plant of the Western Electric Company in Chicago, a Harvard Business School professor named Elton Mayo conducted a series of experiments involving members of the workforce.

His initial goal was straight forward: To determine the extent to which workers and productivity are affected by subtle and not-so-subtle variations in plant lighting, temperature, humidity, rest breaks, and working hours.

The projected results may have seemed self-evident. If workers are tired, surely, they will work more slowly and produce less. If they are allowed short, refreshing breaks at regular intervals, it's reasonable to assume they will work faster and produce more. If the lighting in the plant is too soft or too harsh, they will probably lose their concentration and production will decline. If it's too hot or too cold, the same may happen.

But real life, as Professor Mayo was to discover, is rarely that neat and simple. In the most famous of the experiments, Mayo chose two female workers from the assembly line, and invited them to choose another four.

This small clique of six—all women—were then separated from their fellow workers and placed under the eye of a friendly, personable observer, who would sit at the workshop table with them, listening to their needs and concerns, asking for their advice and opinions, while carefully explaining the nature and purpose of the research.

Then the factory ladies would get down to business, individually assembling an intricate telephone relay device that consisted of some 40 separate parts. This was a thankless, monotonous task, time-tied to the usual strict production targets.

Mayo measured the team's output over several months, making gradual changes to their working routine as he went along. He introduced two five-minute rest breaks, in the morning and afternoon. Production went up. He lengthened the breaks to 10 minutes each. Production rose again.

He switched the workers to six breaks of five minutes each. Production fell slightly. He persuaded the company to provide the women with a free hot meal every day. Production rose slightly. The women were sent home at 4.30 p.m. instead of 5 p.m. Production rose. The women were sent home at 4 p.m. Production remained the same.

Finally, after months of variations, the six women returned to their original routine: a six-day, 48-hour working week, with no rest breaks, no free meals, no early knock-off. Result? In the space of the next three months, the women recorded their highest production levels ever, up 25 percent from the start of the experiment.

To Mayo, none of this seemed to make any sense. No matter the variation in routine or environment, production remained overwhelmingly upward-bound. Then the light of realization dawned. It was the experiment itself that was guiding the process.

Because the six subjects had been hand-selected from the assembly line, because they had been listened to and consulted, because they had been charged with a new sense of individual and collective responsibility, they were made to feel special and significant.

Their laborious work was invested with newfound meaning and purpose. They became a tight-knit team, reacting to the greater freedom and flexibility of their working conditions by spontaneously imposing discipline from within. In the argot of modern motivational theory, we might conclude that the women had been allowed to actualize their true potential and empower themselves.

To put it more simply, they had been made to see that they had the power to make a difference. And, that's what made all the difference (NOSA, 1990).

FOUNDATION OF MOTIVATION

If we apply the lessons of the Hawthorne Effect to near miss incident reporting, we find the very foundation of safety motivation. Near miss incident reporting is not simply a discipline or a set of precepts that must be imposed from above. It must involve the worker, recognize the worker, and perhaps even reward the worker.

It must be a joint effort between management and labor, a pact willingly entered into for the greater good of the company and in the interests of the conservation of vital company assets.

The Ten Rules of Safety Motivation

Enthusiasm

If you are not enthusiastic about safety, how can you expect anybody else to be enthusiastic about it? If you are enthusiastic about the near miss incident system and regard it as a challenge, which can be fun, your enthusiasm will rub off onto other people. Safety management is an ideal opportunity to save a company thousands of dollars due to accidental losses and also an opportunity to show sincere concern for the company's most vital assets, its workers.

The test is: Do you get excited when you tell somebody about the near miss incident reporting system? If not, how can you expect *them* to get excited about it? If so, you are contributing to motivating them to take part in the system. After all, *how many people do you know who became successful at something they hate?* (anonymous)

Avoid Arguments

Experts say that the best way to lose a friend is to win an argument. Avoid arguments about any aspect of safety. If there is a difference of opinion and an argument occurs, a rift, division, or a break in the relationship may result. The best way to win a heated discussion, or difference of opinion on safety, is to avoid it.

People by nature resist wearing protective equipment or improving the house-keeping on a regular basis and any form of argument or difference of opinion may switch the person off. Many near miss systems failed due to arguments over whether it was a near miss incident that was reported or a high risk situation. Don't argue over the detail, accept the report and fix the hazard. That will not lead to deflating of egos.

Never Tell Another Person They Are Wrong

During safety meetings, safety inspections, or any other aspects where safety is communicated, never tell a person that he or she is wrong. This strikes a direct blow to their integrity, judgment, and pride. It may lead to demotivation instead of motivation. If a high risk condition or high risk act is reported as a near miss incident, do not tell the reporter that it is not a near miss incident. That would be telling them they are wrong and possibly discourage further reporting. Remember, there is a small difference between an unsafe situation and a near miss incident that is sometimes hard for workers, and even line management, to determine. Thank them instead, commend them, and act on the information.

When doing an inspection and a deviation is noted, extreme tack must be used to explain this deviation to the person. Never tell a person that his/her safety is poor, wrong, bad, or use any other word that implies that the person's efforts and perceptions of safety are wrong. Remember that negative attracts negative.

For example, on inspecting a workshop, it was obvious that there was very little safety effort. The housekeeping was poor, lighting inadequate, and the ventilation left a lot to be desired. The entire area was dirty and not cared for and was clearly hazardous. The workers in the area also were failing to comply with safety standards.

Thinking carefully, I approached the supervisor of the area and we commenced our inspection. He was waiting for me to tell him how poor his work area was and, thus, strike a blow to his integrity. This would have led to some form of conflict. I noticed he was a rather aggressive person who wanted an argument.

It so happened that I found a small area within the workshop where some individuals had put in an effort to improve the housekeeping. I immediately complimented the supervisor and asked his opinion about the area. My approach obviously caught him unaware. My positive approach was answered by a positive reply from him. This broke the ice. I continued by asking him how he felt about the hazards we noticed, such as oil on the floor, unguarded machines, and rows of completely defective lights. He was now on my side and he proceeded to tell me how poor his workshop was.

He was so used to people telling him that he was wrong that, when I changed my approach, he admitted that the area needed drastic improvement. I did not confront him and, consequently, his pride was not offended. He admitted openly that improvement was needed.

If You Are Wrong, Admit It

Numerous decisions have to be made concerning safety and a lot of activities must be put into motion. If you make a wrong statement, or state a wrong fact, admit it. For example, should you be caught speeding on the freeway, the traffic officer derives great joy hearing all the arguments and excuses he expects from the speeding driver. If you were speeding admit it, take your punishment, learn a lesson, and drive within the acceptable standards.

If you admit that you are wrong, you take the wind out of the other person's sails and there's nothing to argue about. Being mature in safety management is important to create an environment for motivation. If there has been a failure to rectify a reported situation or near miss incident, admit it and commit to time-based actions to fix the situation.

Admitting that you are in the wrong gives you credibility and this credibility is vital when you are attempting to motivate others toward safety.

Always Begin in a Friendly Manner

Always begin in a friendly way when investigating accidents, or near miss incidents, or carrying out other safety activities. Use positive phrases and positive body language. Being open and friendly creates an atmosphere conducive to open heart-to-heart communication for safety.

This advice should be remembered when conducting accident investigations or when interviewing injured victims. In fact, it is applicable during all safety activities. Safety systems often fail because, after an accident, guilty parties are sought. People who happen to be caught committing a high risk act are treated as guilty and punishment and discipline seems to be the only solution. This approach fails. The approach taken in near miss reporting must be positive and could make or break the system and its success.

When approaching others always be pleasant, smile, and find something to compliment. Show an interest in the person. This breaks the ice and builds up trust between you and the other party. Once this trust is established, positive safety discussions can take place.

Use Two-Way Communication

When leading safety committee meetings or presenting toolbox talks, let the other people do most of the talking. Asking leading questions can facilitate this. These entice the others to communicate. Near miss incident recall is an ideal vehicle to initiate this conversation.

Two-way communication is vital in safety. Often safety communication is only in one direction, from the top down, and traditionally little leeway is allowed for two-way communication. Safety communication also can be from the safety advisors to the individuals and should be a two-way dialog. The more people communicate about safety the more involved they feel, which gives them ownership and encourages participation.

When conducting safety inspections, don't always tell the area owner what is wrong. Ask them what is wrong. Ask their opinion, "Do you think that this machine is sufficiently guarded?" "In your opinion, would it help to clean the walkways on a regular basis?" By getting the person to talk, you get their involvement, and you encourage them to give input. Never forget that they are experts within their area. If they make a statement that items need to improve, they are setting their own standards. They are leading the safety drive and do not have to be pushed.

Don't Sell Safety, Let Them Buy

Running around motivating, selling, and pushing safety has little success in the long run. If the leadership forces the reporting of near miss incidents by setting quotas, etc., this will not work successfully. Management needs to apply safety management principles to the best effect to give ownership of reporting and tracking to the employees at the workplace. Make the reporting of near miss incidents interesting and attractive so that employees want to be part of the process and will buy into the system.

By getting employees to buy into the near miss reporting system, we immediately obtain involvement, inspiration, and we impel people to greater safety achievements.

After interviewing hundreds of people over the years, I have yet to find one person who had been "sold" his car, house, TV, etc. All admitted that they bought the items because they wanted them. The salesman was only a facilitator to give the prices, offer discount, and tell about the features.

Apply the same technique to safety. Let employees and line management buy into the system. To do this, one must make safety exciting, a challenge. Competitive goals must be set and monitored on a regular basis. Ongoing feedback and updates on near miss incidents reported keep employees informed and involved and spurs them on to continue observing for hazards.

Have Confidence

To be able to motivate others for safety, you must first of all be motivated and, secondly, must have confidence in yourself and in the system you are promoting. If you do not have confidence in yourself, how can you expect others to have confidence in you. If you do not have confidence in the near miss incident system, you will have difficulty convincing others about the system. Selling safety means that you are selling intangible concepts and benefits, which are sometimes difficult to see and justify. This takes confidence. One must have confidence in the near miss incident system and persevere until it is up and running smoothly. Do not lack faith in the system if it is slow to start or if there is resistance to the program. These are normal reactions to a system that draws all levels around the table and that impacts the very safety culture of the organization.

People love to be associated with, do business with, and to participate with somebody that has confidence. Your confidence will inspire other people to be confident and encourage them to participate in safety.

Reward, Commend, and Compliment

After having asked thousands of people the same question, most admitted that they had recently caught somebody doing something wrong. Either the person committed a high risk act or created a high risk condition or just messed up in general. Asking the same number of people when last they had caught somebody doing something right, the replies were minimal.

We seem to be good at catching people doing things wrong, but very seldom pay attention to or compliment people caught doing things right. In safety, one must make a point of identifying safe behavior. Once safe behavior is identified, some form of compliment, a thank you, or commendation should follow. Any report of a near miss incident or high risk condition or practice should be recognized irrespective of how major or minor the issue reported is. This commending of positive behavior will lead to that behavior being continued.

A lot of successful safety and health systems use tokens of appreciation, which are given out to recognize safe acts and safe conditions. Pens, T-shirts, safety wallets, and other small items may have little monetary value, but have great motivational value.

Health and safety representatives, safety coordinators, and supervisors should be given ownership of this positive behavior reinforcement program. They should have a supply of tokens of "Meal for Two" tickets to award safe behavior. A "Meal for Two" ticket is simply a preprinted voucher that contains the following message:

> Thank you for your excellence in safety. Please enjoy a candlelight supper for two at your local restaurant. (Limited to a reasonable amount.)

This voucher is signed by the responsible person and presented to the person committing the safe act. A candlelight supper for two at the local eatery costs little in comparison to damage, downtime, or injury as a result of an accident and also gets the person's partner involved in safety. This also could be successful as a "Movie for Two" award.

Set the Example

You cannot motivate others toward safety or create an environment in which they are involved, inspired, or impelled toward safety unless you set the example. Always wear the correct personal protective equipment when entering areas that require it. Drive according to the acceptable standards and always adhere to the speed limits. Stop completely at stop signs and red traffic lights all the time. Look after your well-being.

Clean up your garage and carport at home and ensure that when you are working at home that the safety rules applicable at work are also applied at home. If you are a representative of safety, you must walk the talk.

Credibility is most important and is a necessary power base from which to motivate and inspire others toward safety. Becoming personally involved in and participating in the near miss incident reporting and remedy system (NEMIRR) is a must to build credibility for the system and for all involved.

Believe in safety, practice safety, and you will find it easier to inspire others to do the same.

CONCLUSION

The success of any safety management system and, especially the NEMIRR system, relies on the culture within which the system operates. One has to obtain the buy-in from employees and all levels of line management for the system to work.

Implementing a near miss incident reporting system requires a paradigm shift within an organization. This change involves commending employees for reporting events and situations that in the past would have resulted in them receiving some form of punitive discipline. Only by encouraging workers to open up and report near miss incidents can management get a true picture of where the management safety system is failing. As Dan Petersen (1997) said, "An unsafe act, an unsafe condition, an accident are symptoms of something wrong within the management's system."

Motivating to implement and promote safety and health systems is harder than driving safety by force, discipline, and punishment, but is the only way to succeed.

PUSHING THE STRING

One of my mentors laid a length of string on the table and challenged me to move it across the table to the other side. I firstly attempted to push the string, but it merely folded and resisted and did not move at all. I then pulled the end of the string across the table to the other side successfully. "See," said my mentor, "If you push people they shift sideways and resist, but if you lead them, they will follow."

11 Implementing a Near Miss Incident System
Introduction

HOW TO MAKE IT HAPPEN

Business focus has always been on productivity and, in the current environment, it is even more acute. Managers are generally not keen to see workers spending time on nonproductive activities, such as filling in reports or to devote additional resource to investigation and rectification of events that caused no harm.

Certainly, near miss incident reports and the follow-up actions are time consuming and probably add additional expense. The bottom line is that accidents are not good for business, and proven ways to avoid and eliminate accidents, such as acting on near miss incident information, should be the highest priority.

There is a real opportunity for safety improvement by applying a critical level of focus to the near miss incident. Both managers and workers at the coalface (the exposed seam of coal in a mine) need to be encouraged to develop comprehensive systems that can capture, analyze, and rectify close calls and, thus, prevent future accidents occurring.

Near misses need to be recognized as a free, extremely valuable resource in the battle to create an accident-free workplace.

OBSERVATIONS

The near miss incident reporting system is not a system whereby employees observe one another or where management observes employees carrying out their work in order to question the observed actions.

The near miss incident system should not restrict the employees to reporting only near miss incidents. As they are so closely related, reports of high risk work environments and high risk practices should be encouraged as well. Many near miss reports, in fact, will be high risk acts or conditions, but the reporters should not be discouraged from reporting if they are not pure near miss incidents as per definition. Any hazard reported is an opportunity to prevent loss.

To encourage safe work practices, the reporting of "safe work" should form part of the system as well. Safe work is "catching someone doing something right," and is far more powerful in safety than "catching someone doing something wrong." Recognize safe behavior and it will continue.

The opportunity to report safe work would probably be a first for many organizations and would be changing the paradigm where only unsafe work was reported and acted upon. Recognizing one safe act will do more to reduce accidents at the workplace than reacting to hundreds of unsafe behaviors.

RECOGNIZING REPORTED SAFE WORK

Safe work actions and deeds should receive more recognition because they are opportunities to recognize employees' own initiative safety actions. The form of recognition will differ from organization to organization and from safety culture to safety culture. Some institutions issue internal recognition certificates to the employee while others have safety watches that are awarded only on occasion of a big contribution to safety.

There are many ways to take advantage of safe actions reported through the system, but the main issue is taking advantage of "catching someone doing something right" for safety, which is normally contrary to what happened in the past. As stated, near miss reporting systems call for a change in the way safety was viewed in the past.

Having participated in recognition schemes which involved some of the toughest underground miners in the world, I have seen safety recognition change attitudes toward safety. Recognizing employees for an act of safety is a radical safety paradigm shift for the good.

I witnessed the presentation of a buck knife to the recipient of a lucky draw from the month's near miss reports. This was an effort to kick-start the underground near miss incident reporting system that met with a lot of resistance from the "macho miner" culture that prevailed.

All the near miss incident report forms for the month were bundled into a hard hat and a draw was held at the section monthly production meeting. The first form drawn did have the reporter's name and he was presented with a neatly wrapped box containing a buck knife. The atmosphere in the room was electric as he slowly opened the box, unwrapped the tissue, and gingerly removed the shiny buck knife from the wrapping. I could not believe the intense concentration on the faces of these hard miners as they struggled to get a glimpse of the knife. The silence was overwhelming. The recipient rolled the knife over in his hands examining every inch of it before slowly passing it to the person next to him who also lovingly examined the device. The knife was passed on around the table and all the miners admired it lovingly with their eyes while opening and closing the blade. Most of them hunted and fished in their spare time.

The next month the number of near miss reports trebled and not just in quantity, some really high potential events were reported that were rectified before a most likely accident occurred.

An example of "safe work" reported reads as follows: "Tom Harper replaced the machine guard that contractors had left off of pump #43 at the entrance gate at Plant 5."

SAFETY SUGGESTIONS AND RECOMMENDATIONS

Often safety suggestions or recommendations find their way into the near miss reporting system, which is another bonus of the system. These should be reviewed and the necessary action taken and feedback given as if the report was of a near miss incident. One such suggestion received by an organization stated, "Suggest all meetings begin with the fire drill and emergency procedure for that room."

CONSTRAINTS

Legal concerns may compel an organization to analyze an accident thoroughly, but may also inhibit the use of near miss incident data. For example, showing that an organization knew about a particular precursor, but did not take corrective action could increase the organization's liability in the event of an actual accident. As a result, some organizations may be reluctant to establish near miss incident reporting systems and they may rely on oral, rather than written, notification of observations.

A counter argument is: Which is the lesser of the two evils? Failing to identify a hazard *or* identifying a hazard and not eliminating it? Identifying hazards is more proactive and shows good intent and should receive more favorable consideration than having no system to report and identify hazards. As my Australian colleague put it, "You're buggered if you do, and buggered if you don't."

UNDERSTANDING

The first step in implementing a near miss incident recognition, reporting, recording, and remedy (NEMIRR) system is the understanding of the significance of near miss incidents. The leadership of the organization must have a clear understanding of the definition of a near miss incident. The union and the workforce must all understand the philosophy and benefit of reporting these events even though "nothing happened." This may involve selling the concept to all levels within the organization and holding training sessions, or including the topic in toolbox talks and safety meetings.

Management should develop a written standard pertaining to near miss incident reporting and this could be used as a training outline for creating understanding of the importance of these events and the role all must play in the system. Only once all within the organization are aware of the difference between a loss-producing accident, a property-damage accident, and a near miss incident can the program be rolled out.

BENEFITS

The benefits of implementing and maintaining a NEMIRR system must be understood and be made known to the members of the organization.

The main benefit is the reduction in the number of accidents that cause injury to employees and damage to property. Another major benefit of near miss reporting is that it is easier to get to the root causes of the event since nobody has been injured or killed, so there is no pressing need for a cover-up. According to Jones et al. (1999), case studies of offshore oil rigs have indicated a 60-percent reduction in disabling injuries after the 10-fold increase in near miss incident reporting. Other reports indicate that there is a definite correlation between the number of near miss incidents reported and the reduction in the number of serious injuries experienced. The results were only achievable by following up on the reports and after risk-ranking them, identifying, and eliminating the root causes.

Because near miss incidents are the foundation of a major injury, their identification and elimination will bring about a reduction in the number of injury accidents.

A third very important benefit is the reporting and recognition of safe work and safe behavior. This is perhaps a new concept to the organization and many line managers and supervisors will have difficulty implementing this new and different approach to safety. The NEMIRR system allows for the reporting of high risk situations, but also situations where employees deserve recognition for following the safety rules or otherwise contributing to the safety program. This aspect of the system is difficult to start, but once in motion will break the traditional ice between supervision and workers.

As James P. Bagian (2004) so aptly reported:

> Another important decision is what should be reported. Is the purpose of the reporting system to look only at things that have caused an undesirable outcome, or is it also to scrutinize other things, such as close calls that almost resulted in undesirable outcomes but did not, either because of a last-minute "good catch" or because of plain good fortune? Close calls are extremely important areas of study because they are much more common than actual bad outcomes. Thus, close calls provide repeated opportunities to learn without first having to experience a tragic outcome. In addition, because close calls do not result in harm, people are generally more willing to discuss them openly and candidly, because they are less fearful of retribution for the part they played in the event. Also, people are often more motivated to analyze close calls if understanding them is considered an opportunity to act proactively to prevent undesirable outcomes in the future (p. 39).

BUY-IN

The NEMIRR system cannot and should not be the sole responsibility of the safety department. This system involves all employees at all levels and calls for total participation or else the system will fail. Buy-in from all levels must be obtained before launching the program or resistance will block all efforts to run a working system. Management at all levels, unions, union members, and employees must all see the benefit of the program and must be willing participants. Many will be skeptical and will remain so about the system until they see others participate and the system functioning effectively.

REPORTING

Near miss incidents cannot be assessed or managed until they are reported. The finest mechanism for reporting near miss incidents is the workforce. The workforce must be encouraged and enthused to report near miss incidents and should not be penalized when they do so.

Some of the main obstacles to the reporting of near miss incidents include:

- The employee is held accountable for the near miss incident.
- The employee is challenged with, "What have you done about it?"

Date / Form #	Originator/ Report Type	Safety Issue Description	Status	Complete
9/8/10 000038	Jim Brown Unsafe Condition	Weed control — excessive weeds and trash around compound at Plant. Potential insect, animal bites as weeds are taking over our equipment.	Sent e-mail to maintenance on 9/8/10	
9/8/10 000039	Adrian Unsafe Condition	Noted and corrected 5 trip hazards at Cl2 Containment Bldg.	Sent thanks to Adrian 9/8/10	Completed 8/30/10
9/8/10 000040	Bill Near Miss	Tripped and nearly fell. C/2 power supply(GFI box) is a trip hazard as it stands about 20 inches above ground.	Mike will move power supply.	
9/8/10 000041	Wally Unsafe Condition	Booster # 4 wire size to 30 HP motor is undersized.	Note to Mitch on 9/8/10	
9/8/00 000042	Randy Unsafe Condition	Booster # 5 wire size to 40 HP motor is undersized.	Note to Maintenance on 9/8/10	
9/8/10 000043	Shawn Unsafe Condition	Unprotected live electrical surfaces in main panel exposing maintenance electricians to serious dangers. Also, unsafe dual control switches for motor operation exposing equipment and personnel to potential dangers.	Work Order to electrical dept. on 9/8/10	
9/8/10 000044	Jennifer Unsafe Condition	Sodium light blinking on and off at NE corner of Building. Area is dark.	Work Order to electrical on 9/8/10	
9/8/10 000045	Charles Near Miss	All wheel nuts came loose on driver's side rear tire — almost lost tire on vehicle #1197. Fleet Services responded.	Thanks sent to Charles on 9/8/10	Completed 9/8/10

MODEL 11.1   Typical examples of reports from near miss incident reporting system.

- The employee is punished.
- Embarrassment.
- Macho attitude.
- Employees see no action taken on the report.
- Employees do not receive recognition for reporting.

The above mentioned are some reasons why near miss incident reporting systems do not work.

I have worked with organizations where between 700 and 800 near miss incident reports were received and which were reported by hourly employees, middle management, and senior management every month. Everybody was involved in the near miss incident reporting system. What made it even better was that each near miss incident reported had already gone through a risk assessment process by the person doing the reporting (Model 11.1).

NO NAMES

The trick to the success of a near miss incident reporting system is to make it a "no names" reporting system. Discipline must be entirely removed from the system and employees must feel comfortable to report these undesired events. This may call for a radical paradigm shift within the organization. If a near miss incident, with the potential to kill somebody, is reported, no punitive action should be taken. This will deter people from reporting near miss incidents. An organization can only learn from its mistakes if they are revealed.

To make safety work, we should play the ball and not the man. A safe space must be created for reporting. Once the right atmosphere is created, near miss incident reporting will occur and slowly become the normal thing to do.

RESISTANCE TO CHANGE

The introduction of a "new" safety and health system or element of a safety management system, such as near miss incident reporting, will inevitably be met with resistance from various levels and individuals within the organization. This is normal and expected as with any new safety idea, program, or drive.

There are four main reasons for resistance to change.

- The first is that resistance occurs because it threatens the status quo or increases fear and the anxiety of real or imagined consequences. "Will I get into trouble for reporting near misses? Will I be held accountable if someone reports a near miss incident in which I was involved?" This includes threats to personal security and confidence in an ability to perform. It demands a shift from one's comfort zone, a change of paradigm.
- Change also may be resisted because it threatens the way people make sense of the world, calling into question their values and rationality and prompting some form of self justification or defensive reasoning, such as: "We've tried that before and it didn't work."
- Resistance may occur when people distrust the intent of the change or have past resentments toward those leading change. Will this scheme be used by management for discipline? Change is also disruptive and is inclined to "rock the boat" for some.
- Introducing a new way to practice safety could affect employees and some levels of management if they have different understandings or assessments of the situation. They may be in a safety comfort zone that they feel will be threatened.

BARRIERS TO REPORTING

Why is it so difficult to get employees to report near miss incidents? They could be discouraged by one of these common barriers:

- Employees don't know they are supposed to report near miss incidents, after all, nothing happened.
- They don't know how to go about it. The training was insufficient or the report methods are not clear.
- They are afraid of being reprimanded or disciplined for actions that led to the near miss incident.

- Employees feel pressure from co-workers to keep quiet so that nobody gets into trouble and nobody loses the safety bonus or spoils the "safety record."
- They are under pressure to maintain a clean incident/accident/injury record so that the team will get the safety bonus or reward.
- They are new to the crew and want to make a good impression. Why make waves and stand out in the crowd all for nothing?
- The work culture says, "Nothing happened, no one was hurt so don't make a big deal out of it."
- Co-workers view the near miss incident with humor instead of seeing the hazard. If everyone is laughing, how serious could it be?
- Last time they tried to talk to the supervisor about a near miss incident they were belittled or disregarded.
- It's just too much trouble filling out those forms and they have no time for paperwork.
- Employees, in general, dislike paperwork.
- "We tried other safety things in the past and they didn't work."
- Near miss incident reporting is not encouraged by the organization.
- Language barriers.

Bridges (2000) focuses on barriers that inhibit disclosure. To increase report rates, he advocates that nine barriers be overcome. These include:

- Potential recriminations for reporting (fear of disciplinary action, fear of peer teasing, and investigation involvement concern).
- Motivational issues (lack of incentive and management discouraging near miss reports).
- Lack of management commitment (sporadic emphasis, and management fear of liability).
- Individual confusion (confusion as to what constitutes a near miss and how it should be reported) (p. 379).

LONG-TIME EMPLOYEE

A long-time employee once told me that he would never ever report a near miss incident again. When I asked him why, he told me that the last time he reported a near miss incident to his supervisor, he was sent home for three days with no pay. "I'll never report one again," he confirmed.

These are a few of the reasons why a near miss incident quickly vaporizes away and doesn't get captured. The most common barrier to reporting of near miss incidents is the perception that "nothing happened," and the other all-time challenge to the leadership, "Nothing will be done, so why waste time to report it?"

12 Implementing a Near Miss Incident Reporting System
Implementation

SETTING THE STANDARD

A management decision in the form of a procedure, operating policy, or standard on near miss incident reporting must be issued. This will become the guiding document that sets the foundation and guidelines for the system.

The National Institute of Occupational Safety and Health (NIOSH) lists "management commitment" as the first of four key elements needed for safety performance. Standards are written commitments.

Standards should exist for all elements of the safety and health system. They should define what the standard should be, who is responsible for certain actions, and by when these actions should take place.

POLICY

An example of a near miss incident reporting and recording system policy reads as follows:

> At our company, we are committed to the implementation and maintenance of a near miss reporting/rectification system so that everyone can identify and report near misses, unsafe acts/conditions, rank the potential, and prescribe corrective action to reduce risk. By getting to the near miss incidents, potential losses can be prevented before rather than after the event.
>
> At present, we don't encourage employees to report near misses and we tend to be reactive, in our safety activities. The system will put us into a proactive rather than a reactive safety mode, and help us gain control over potential areas of loss.
>
> We can't be injury free unless we are near miss incident free and until we identify, report, and rectify near miss events, or the accidents we haven't had yet.
>
> Our near miss reporting system will create employee involvement in identifying all forms of risk and help us recognize potential losses. It will help us to get to zero injuries and zero citations, and we will appoint, involve, and encourage 750 Safety Observers (our entire workforce) in the workplace putting us into world's best practice and a leader in the ranks.

In order to encourage near miss incident reporting, the following paragraph was included in the company's safety newsletter:

> Letting a near miss incident go unreported provides an opportunity for a serious accident to occur. Correcting these actions or conditions will enhance the safety within your facility and provide a better working environment for everyone involved. Don't let yourself or co-workers become statistics—report near miss incidents to your supervisor. Prevent an accident that's about to happen!

STANDARD

The following document is an example of an organization's near miss incident system standard that describes the organization's commitment to near miss incident reporting and investigation, and also clearly spells out responsibilities.

Objective

The objective is to define the methodology for reporting and investigating **noninjury (loss-producing) accidents** and **near misses** so that the immediate and basic causes of the events are identified and recommendations to prevent a recurrence are proposed and implemented.

References
- (References to local safety legislation)
- (References to existing safety program elements)

Definitions

Accident: An undesired event that causes harm (injury or ill health) to people, damage to property, or loss to the process (production or business interruption). (This includes fires, as there is a loss.)

Near Miss: An undesired event that, under slightly different circumstances, *could have* caused harm (injury or ill health) to people, damage to property, or loss to the process (production or business interruption). (There is no loss.)

Serious Reportable Accident: Reference to local legislation, e.g., fire, landslide, or explosion resulting in losses to production or to production equipment.

Responsible Person (RP): This is the manager/superintendent/supervisor of the area in which the event occurred, or the most senior person on the site where the event occurred.

Near Miss Accident Investigation Form (NMAIF): The company NMAIF is to be used for the investigation of all high potential, near misses, property damage, and injury producing accidents.

Standard
- All accidents that result in injury or damage, and all near misses, shall be reported promptly so that an investigation can be launched to determine the root causes, so that corrective action can be taken to prevent recurrence.

- The Near Miss Accident **Initial Report** form is to be posted on the e-mail system before the end of the shift by the **Responsible Person (RP)** in the department that experienced the event.
- The company Near Miss Accident **Investigation Form (NMAIF)** is to be used for the investigation of all high-potential (use risk matrix on form) near misses, property damage, and injury producing accidents.
- A risk assessment of the event must be indicated on the risk matrix included in the form.
- For all accidents involving cars/pickups/trucks on company property, in addition to the abovementioned form, a separate form, the company **Vehicle Accident/Damage Report** must be completed by company security because without this form outside repairs cannot be done (see Company Element: Near Miss Accident Investigation).
  - Accidents that have high potential for loss should be thoroughly investigated.
- Results of investigations shall be shared with others via five-minute talks, SHE (Safety, Health, and Environment) Committee meetings and similar avenues, as appropriate.

Procedure and Responsibilities (Property Damage/ Environmental/Vehicle Accidents/Near Miss)

Any employee involved in, or who witnesses, a damage accident, near miss, however trivial, shall:

- Notify his supervisor immediately.
- Arrange to make the scene of the accident safe and ensure that site evidence is not destroyed unless unavoidable to prevent further injury or damage.
- Cooperate with the investigation.

The **Responsible Person (RP),** upon being notified of the property damage or near miss, shall (where applicable):

- Visit the scene of the event.
- Arrange to make the scene of the accident safe and ensure that site evidence is not destroyed unless unavoidable to prevent further damage or injury.
- Obtain written statements from the witnesses as soon after the event as possible, but preferably during the same shift.
- Publish the Near Miss/Accident Initial Report before the end of the shift.
- Commence the accident/near miss investigation process and head the investigation meeting.
- Complete page 1 of the near miss/accident investigation form within 24 hours and submit to the SHE Department.
- Finalize the investigation as soon as possible and implement the remedial measures immediately.

- Circulate the investigation findings to all sections within the department and the rest of the plant for their information.
- Ensure that both the immediate and root causes of the event are identified.
- Follow up to ensure that the remedial measures have been implemented as soon as is practicable after the event.
- Ensure that the investigation form is completed correctly and submitted to the next level of authority (one-up manager) for signature.

Investigation and Reporting Requirements

Near misses and damage accidents involving company activities include:

- High potential near misses and accidents causing damage shall be reported and thoroughly investigated using the NMAI Form.
- The **Responsible Person (RP)** will nominate an investigator for the particular event (nominated accident investigator [NAI]) and commence the investigation with him/her. The NAI should not always be the SHE superintendent.
- The RP also will initiate and circulate the Near Miss/Accident **Initial (NMAI) Report** within the same shift, even if digital pictures are not available in time.
- The front page of the NMAI form is to be completed and a copy sent to the SHE department within 24 hours by the RP.
- The safety superintendent (SAS) (SHE department) will record the event in the register and allocate a tracking number for follow-up purposes.
- The company booklet, *Accident Investigation,* can be used as a guide.
- During normal working hours, the investigation shall be started as soon as possible after the occurrence and completed within 72 hours. If more time is required, notify the central register holder (SAS).
- Outside of normal working hours, the investigation shall be initiated by the RP in whose area of responsibility the event occurred and a NAI shall be nominated.
- The NAI shall record findings in the Near Miss/Accident Investigation Form (NMAIF) and both the immediate and basic causes of the event must be determined. The estimated costs of the losses should be entered on the form.
- The investigation shall include recommendations for actions to prevent recurrence, and list:
  - WHAT should be done to prevent a recurrence.
  - WHO is responsible for doing the work/taking action.
  - BY WHEN the actions are to be completed.
- Only once these actions have been implemented, should the NAI and RP and the SHE superintendent sign the form.
- The signed form must now be circulated to the next level of supervision for comments and signature.

- Managers are accountable for ensuring that the immediate and root causes of the near miss/accident has been identified and that they have been eliminated or mitigated and should only sign the form once this has been done.
- Once signed by the manager, all high potential near misses, damage/interruption investigations are to be circulated to the general manager concerned for comments and signature.
- Once signed by the general manager, the form is returned to the SHE Department for filing and sign-off in the register.
- The completed investigation report shall be issued within 72 hours of the event.
- Advice and assistance in investigation of high potential near misses and property damage cases can be obtained from the departmental SHE superintendent or SHE Department.

The requirement for near misses/accidents involving contractors is that the company person responsible for contractors (to whom the contractors report) in each area (i.e., directly supervised or those on a fixed contract) shall ensure that appropriate reporting and investigation of near misses and damage/interruption accidents is carried out according to this standard as if the contractor was a company employee.

AMNESTY

If management wants this system to work and contribute to the reduction of risk and consequent losses to the organization, a decision to declare amnesty must be made. The reporting mechanism should allow for anonymity of reporting, but also allow for the reporter to volunteer his/her name if he/she feels comfortable to do so, such as in the case of reporting safe work or deeds. Blame-fixing and punitive actions based on reports where the employee was obviously at fault must be avoided. This may be a difficult paradigm for most managers to change as most link safety violations with punitive actions.

The employee grapevine works well and the first time there is punishment leveled at an employee who reports a high risk act of another employee, or themselves, the reporting will dry up. Management must make a bold decision in this regard and accept that discipline in safety has never worked and never will.

As the safety icon Frank E. Bird, Jr. (1996) said:

Punishment is the last resort, but it must be done in a way that communicates your genuine concern (p. 52).

Frank Salas, Joint Union Management Safety Coordinator, once told me: "Discipline in safety hasn't worked for 20 years, why will it work now?"

If an organization is running its safety program on discipline and still experiences a single injury, the approach isn't working and it's time to try something different.

CREDIBILITY

The near miss incident system should not be seen and used as a fix-all for safety system failures and resultant losses. It is not a stand-alone, "silver bullet" safety program and will never be. It forms a part of the entire safety management system and should be viewed as one of the critical elements of a 60- to 70-element safety system.

NIOSH recommends than an organization establishes and communicates a clear goal and objective for the safety and health program and to provide visible top management involvement in implementing the program.

For the system to be credible and acceptable to employees at all levels, it must have the support of the leadership of the organization. There should be a commitment from the executives to participate in the system and to maintain it as an ongoing process rather than just a "flavor of the month" safety gimmick. Feedback to near miss incident reporters and the workforce and follow up action on hazards reported is essential to give the system credibility.

According to Mike Williamsen (2009), there are many factors needed in order for near miss incident reporting to be effective. Included are a willingness by the organization to learn from its mistakes, open and honest communication between workers and supervisors, and the removal of fear of reprisal for reporting, positive motivation to report near miss incident situations, and positive actions taken immediately to eliminate the problems reported.

One of the main reasons for the failure of a near miss incident reporting system is the lack of quick results. The system must be allowed to grow and develop. An instant flood of near miss incident reports should not be expected.

WHAT ARE CRITICAL SAFETY AND HEALTH SYSTEM ELEMENTS?

Critical safety elements are elements that include environmental and employee factors and which need to be controlled constantly to prevent losses occurring from, for example, injuries, disease, damage, production delays, and environmental harm. These elements form the ongoing safety program or, more aptly termed, *safety and health management system*.

Critical safety elements are those elements most likely to give rise to losses. Past experience based on thousands of safety inspections and audits have shown that control over certain aspects of the environment and work practices, can significantly reduce accidents.

Controlling critical safety elements is precontact safety control. It is effort directed toward the prevention of undesired events. Once controls over certain critical elements are exercised, proactive safety is practiced.

EXAMPLES

Controlling the stacking and storage procedures in the workplace is an example of control over a critical safety element. Ongoing housekeeping campaigns, inspections, audits, and permits are other examples. A safety program is designed to exercise control over potential areas of loss. All critical safety elements are potential areas of loss. Near misses are visual warnings of potential loss.

PRINCIPLE OF THE CRITICAL FEW

The safety management principle of the critical few states that "a small number of basic causes could give rise to the majority of safety problems." A few critical jobs could be responsible for the majority of injury-producing accidents occurring and these few critical items (critical safety elements) should receive maximum safety control to minimize their potential for causing (the majority) of problems.

Precontact control is proactive and directs the safety efforts toward focusing on these crucial areas before a loss occurs. Most safety programs are reactive and only institute controls after an accident has occurred. This is called postcontact control, (fire fighting, patch prevention), or treating the symptom and not the cause. Near miss incidents indicate the causes of system failure and, therefore, play a vital role as a critical element in a structured safety system (program).

WHY THESE ELEMENTS?

Experience has shown that there are between 60 and 80 critical safety arenas (elements) that must be controlled constantly to constitute an effective safety program. Many are dictated by local and national safety laws and legislation and are compulsory anyway. These legal requirements should be viewed as the minimum standard and management should extend the safety program to best practice rather than only complying to the minimum legal requirements.

These elements may vary from organization to organization and from industry to industry. The emphasis on individual elements also will vary according to the nature of the process, risks arising, culture of the workforce, and category of business, such as mining, the iron and steel industry, transportation, the fishing industry, manufacturing, etc.

BENEFIT

The benefit of controlling critical safety elements is that the work being done to manage safety is channeled at reducing the risk and potential loss in areas that have been identified as crucial. Some critical safety elements help control the physical conditions that would contribute to the reduction of losses as a result of an unsafe work environment. Other critical safety elements are directed toward the control of the persons within a workplace. These controls would include items such as critical task procedures, rules, training, and activities to involve, motivate, guide, and train employees in safe work practices.

ENVIRONMENTAL AND BEHAVIOR

Numerous safety program elements help control, both the behavior of people at work as well as the work environment and the work procedures. No hard and fast dividing line can be drawn between elements defining them as either behavior or environmental control. The one influences the other. Near miss incident reporting and rectification systems could cover both environmental and behavioral aspects.

Elements to control the environment include items such as:

- Housekeeping
- Lighting
- Electrical safety
- Stacking and storage
- Ventilation
- Safety inspections
- Lifting gear safety
- Demarcation of walkways
- Machine guarding
- Hazardous substance control

Elements to control behaviors to ensure safe work procedures could include:

- Critical task procedures
- Rules and regulations
- Permit systems
- Safety and health training
- Appointment of a health and safety representative
- Holding regular safety meetings
- Safety communication sessions
- Near miss incident and accident recall
- Safety promotion
- Pre-employment medical examinations

SAFETY AND HEALTH MANAGEMENT SYSTEM (PROGRAM)

When asked what safety management system (program) they were following, numerous safety practitioners and safety coordinators were at a loss for an answer. Complying with the legal safety requirements is the minimum standard, but employers should strive for best practice in safety and institute a safety management system. This system should be risk-based, management-led, and audit-driven.

DEFINITION

A Safety Management System (SMS), traditionally termed a safety program, is the driver of a company's safety culture and is defined as "ongoing activities, processes, and efforts directed to control accidental losses by monitoring critical safety elements on an ongoing basis to reduce safety and health risks."

The monitoring includes the promotion, improvement, and auditing of the critical elements on a regular basis.

EXAMPLES OF AN SMS

An excellent guideline for a structured safety and health program is the U.S. Occupational Safety and Health Administration (OSHA) Volunteer Protection Program (VPP). Excellent safety program outlines also are given in the American National Standards Institute (ANSI) standard: ANSI/AIHA Z 10 – 2005, American National Standard—Occupational Health and Safety Management Systems.

Other internationally recognized system guidelines to comprehensive safety programs, such as OHSAS 18001 and the ISO series, have already been mentioned in this book.

INFORMATION

Information about the NEMIRR system must be made known to all. Internal communication systems can be used to this end, and the intranet, newsletters, and other internal communication methods can be used as well. If possible, the reporting forms can be made available online at computer terminals and feedback on the systems success can be broadcast on the internal site.

THE FORM OR BOOKLET

The reporting and recording form should be as simple as possible, be portable, and always available. Bulky reporting sheets that cannot be carried by the employees will not work. A small, pocket-size card or booklet is ideal and has proved to be very successful in a number of instances (Model 12.1).

Depending on the sophistication of the workforce, a risk matrix should be incorporated into the reporting form/booklet so that, once trained, the employee can do a risk ranking of the near miss incident as it is reported.

The risk ranking will indicate whether or not the event warrants a full investigation, as would an injury-producing accident. This aspect of the program is crucial because all near miss incidents with high loss potential and high probability of recurrence are the events that may have ended up as loss-producing accidents, had it not been for the slightly difference circumstances or conditions. This means that the safety efforts can now be levered at those near miss events that had a high likelihood of being serious accidents.

Events reported that have lower risk rankings will be items that the reporter rectified themselves or near miss incidents that warrant attention, but not necessarily a full blown investigation (Model 12.2).

NEMIRR TRAINING

When we introduced the system at a large copper mine and smelter, we decided to do things differently. Our pocket size, perforated, and self-carbonized booklets were ideal for underground miners to slip into their pockets where they could stay dry and safe in wet underground conditions. They were also simple, easy to complete forms, and the reporter could retain his carbon copy in the booklet.

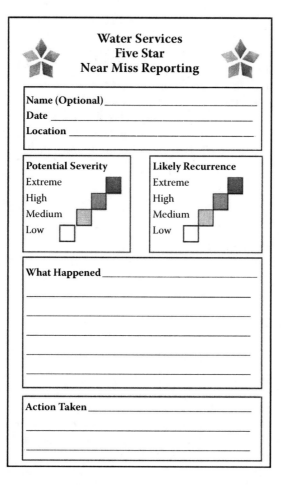

MODEL 12.1 The PWSD near miss reporting form. (From Phoenix Water Services Department's (PWSD) *Near Miss Reporting/Near Miss Rankings*. 2010. Phoenix, AZ. With permission.)

Some miners saw the booklets while they were being delivered from the print shop and excitingly demanded copies. Seeing the opportunity to "sell the system," we told them that they first had to attend the near miss training session before they could get a booklet. Their reaction was typical: "You safety guys load us with tons and tons of paperwork that we don't want and now we ask for safety paperwork and you won't give it to us!"

The four-hour sessions on the theory of near miss incidents and on how to use the booklet to report near miss incidents were scheduled and were oversubscribed for the first two weeks. We had to turn employees away on a number of occasions as the training room was full. As the employee/management safety representative, Frank Salas said,

> We've always had to drag the guys into safety training classes and now we have to turn them away! This is the first time in 30 years that that has ever happened!

Risk Ranking	
Possibility of Reoccurrence	
LOW	Potential for reoccurrence one in 20 years
MEDIUM	Potential for reoccurrence annually
HIGH	Potential for reoccurrence daily
EXTREME	Potential for multiple times per day or each time the task is performed
Loss Potential	
LOW	Loss will be minor, property damage slight, and first aid only
MEDIUM	Property damage less than $10,000 or lost time less than 4 days
HIGH	Property damage greater than $10,000 or lost time greater than 4 days
SEVERE	Property damage greater than $50,000, multiple employees injured (lost time), or death

MODEL 12.2 The PWSD risk ranking. (From Phoenix Water Services Department's (PWSD) *Near Miss Reporting/Near Miss Rankings*. 2011. Phoenix, AZ. With permission.)

Our approach created a value system around near miss reporting and, at its peak, we received 2,212 written reports in one month from a workforce of 2,600 employees.

Training Outline

To create understanding concerning near miss incidents and the reporting system, a brief training session outline should cover the following:

1. High risk acts and conditions
2. Definition of an accident
3. Definition of near miss incidents
4. The accident sequence
5. Near miss incident/accident root causes
6. Exchanges of energy
7. Explaining losses after an accident
8. Property damage accidents
9. The accident ratio
10. What determines the difference?
11. The luck factors
12. Close calls, near miss incidents
13. Examples of near miss incidents
14. Risk ranking of near miss incidents
15. The importance of reporting
16. How to report (filling in the form)
17. What if I don't have a form with me?
18. Incident and accident recall
19. Verbal reporting
20. No names required
21. Benefits of reporting

The training should include a practical exercise where examples are used and the attendees go through the process of actually filling in the form and practice ranking the risk of the near miss. This will make them more familiar and comfortable with the reporting process.

FORMAL REPORTING

A formal reporting system is where the observer fills in the report form and posts it in the near miss box or hands it in to the supervisor who then enters it into the system. The formal system is the main reporting method, but requires access to a form, or a computer terminal, or a "safety reporting hotline."

SAFETY REPORTING HOTLINE

A good idea to encourage the reporting of high potential near misses and other safety issues that require urgent attention is the installation of a dedicated internal telephone number as a safety hotline. This means any employee can use the internal phone system to report near miss incidents or other safety-related matters. The caller remains anonymous and does not have to identify himself to the operator on the other end of the line. No one in the immediate vicinity knows what number they are calling or who they are speaking to and this gives the reporter confidence to report without the chance of him or her being ridiculed or teased by other colleagues.

INFORMAL REPORTING

Informal reporting is when near miss incidents are reported verbally without filling in the form or online report. This could be an employee who tells his supervisor about a safety deviation. It could involve an employee telling another employee about a certain situation. Observations discussed at safety meetings and tailgate safety sessions would fall into the informal category.

All safety meetings and pretask talks should begin with an incident recall session as part of the informal reporting process. The discussions from incident and accident recall sessions are informal reports and form an invaluable part of the NEMIRR reporting system.

RISK RANKING

Risk ranking of the event reported may deter reporters from reporting and participating in the system if the procedure is too complex and they do not understand it fully. The best advice here is to keep it very simple. During the training session, the attendees should go through the motions of actually ranking near miss incidents examples so that they are familiar with the process. They should be taught how to use the simple risk matrix and should understand that risk assessment is not a perfect science and that we all see risk differently. There should be no concern if different attendees give different rankings for the same hazard. People see things differently and have different perceptions of risk.

It is important to emphasize that risks be ranked on the "normal expected loss" approach and not the worst case or best case scenario. Often, emotions slant the rankings "because it's a safety issue."

INCENTIVES

The near miss reporting system does not rely on incentives, reward, or gimmicks for reporting or for the success of the system. The organization may want to introduce some form of recognition system for reporting, but many of these systems are viewed with suspicion by the workforce. The worst system is one that offers awards from a glossy catalog if an employee is "injury free" for a certain period. This approach has unfortunately spoiled the safety incentive system as employees simply do not report the injury to be able to take home the fancy prizes to their families.

The simplest and best reward system for reporting is a letter (or e-mail) from line management acknowledging the near miss report and indicating what action they intend making on the reported matter. The finest reward to employees is the fact that their reported issue has been acknowledged, addressed, and fixed. That's what gives them recognition and power.

COLLECTION SYSTEM

A central collection area should be established where all reports are collected. This is normally an administrative function even though, in some organizations, a safety coordinator is allocated the responsibility of administrating the collection, analysis, and submission of near miss incident reports. Both formal and informal incident reports should be collected and tabulated on a simple spreadsheet.

The spreadsheet should list the report number, the area concerned, a description of the situation, the risk ranking of the event, and corrective action required. If action has already been taken, this also should be listed. Pending actions should be timelined.

FEEDBACK ON REPORTS

It is vital that some form of feedback and follow up on the reported near miss incidents is given. Generally, a monthly list of all the items reported is circulated to all via the intranet or posted on the notice boards, or featured in the newsletter. Employees want to see what they have reported in writing and want to see some form of action being taken on their report. In most instances, the reports will be of high risk acts or high risk conditions and safe work observations and few may be true near miss incidents.

What is important is the feedback and action instigated by management to rectify the hazard if the event reported wasn't already closed out at the time of the report. For example, an employee notices some 2 × 4s with protruding nails laying in the walkway. He removes the nails and places the wood in a proper storage area. He then fills in the near miss incident report form as follows:

Describe the near miss incident: "Nearly stepped on four 2 × 4s with nails protruding in recycle area walkway." Action taken to prevent recurrence: "Removed nails, restacked the timber, and mentioned it to the area supervisor."

In this case, the event loop is closed, but nevertheless the reporting employee wants to see his report published on the monthly summary sheet. The feedback given by the publishing of the report reinforces his or her faith in the system and he/she is spurred on to continue reporting. He/she is now participating in the system and involved in the safety process. The management principle of *safety participation* states that safety motivation increases in proportion to the amount of participation of the employees involved. To sustain this participation, there must be feedback on each and every event reported whether they are pure near miss incidents or just high risk acts and high risk conditions.

13 Implementing a Near Miss Incident Reporting System
Follow Up

INTRODUCTION

For the system to succeed, it is important to note that all the events reported cannot, and should not, be subject to a full near miss investigation process. This will bog the investigators down and prove too time consuming.

In the introductory stages of the system, it is recommended that the high/high ranked events be investigated first. These high risk events should be investigated with the same rigor as injury-causing accidents.

Once the system has matured, the investigation can be extended to the high/medium ranked near miss incidents and other reports that warrant investigation. Although theory dictates that all deviations should be investigated, in practice, this cannot happen and focus should be on the high risk situations reported. Admittedly some form of follow up and review of all events must occur, but not necessarily a full investigation.

Many of the reported events already will have been rectified so no further action will be required until a trend is identified that may indicate the need for further investigation. A number of near miss incidents reported involving vehicles may be an indicator that the transport safety program needs review and steps need to be taken to improve drivers' skills/habits, etc.

REPORTING SYSTEM FOLLOW UP

INVESTIGATION

High risk potential near miss incidents and other high risk situations, such as high risk practices and high risk work environments, should be investigated. The purpose of the investigation is to find the immediate and root causes and then eliminate the problem. Remember to fix the workplace and not the worker.

Investigation of high risk near miss incidents offers an opportunity to investigate an accident that has not yet occurred and should be viewed with the same importance as an injury-producing accident.

REMEDY

Once the investigation is completed, the hazards need to be eliminated. If the loop is not closed, the same risks will eventually lead to a loss-producing event. The hazard must be eliminated and a follow up done to ensure the remedial action is completed. This is a weak area in many near miss incident and accident investigation systems.

ALLOCATION OF RESPONSIBILITY

Responsibility for corrective actions must be delegated to a Responsible Person (RP), work team, or department, and a completion date set. All actions must be time-lined and should not be left open-ended. If not, they will never be completed. Actions must be positive. For example, "I will tell employees to be more careful in future," is not a positive action. "We will hold a five-minute team briefing on the importance of eye protection and reissue safety glasses to all" is more positive.

CENTRAL RECORD AND PUBLICATION

A central collection point for all near miss incident reports should be established. From this, base reports can be summarized, analyzed, and sent to the correct departments for corrective action. Closed-out reports can be added to the central register.

The central register should be displayed on the internal network or circulated in hard copy on a weekly basis. Many organizations publish the report in the monthly newsletter. It is important that the information be publicized because this is the key to the success of the system. Employees can see their safety inputs in print and feel they are empowered in the safety process.

A GOOD EXAMPLE

A large underground mine had a write-on board erected in the area where the miners collected their cap lamps at the beginning of their shift. The board listed all the underground near miss incident reports and a separate column on the board showed what action had been taken for each reported event as well as the completed dates. Miners could follow the progress of their reports each day because of this system.

FOLLOW UP AND CLOSE OUT

Outstanding remedial actions should be kept on the register until completed. Closed out actions can be removed from the system as it is updated.

STATISTICS

As part of the normal injury statistics that are published by the organization, near miss incident statistics should be tracked and published as frequently as the other safety statistics. Statistics published should include:

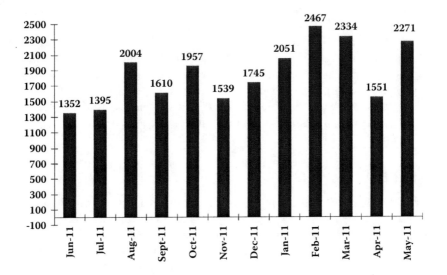

MODEL 13.1   Number of near miss incidents reported.

Area/Division	Near Miss Reports	Number Complete	Percentage Complete
Area 1	1231	1231	100%
Area 2	494	494	100%
Area 3	111	111	100%
Warehouse	163	163	100%
Maintenance Services	54	50	92%
All Divisions	2053	2049	99%

MODEL 13.2   Near miss incidents rectified.

- Number of near miss incident reports received during the period
- Near miss incident reports per area
- High potential near miss incidents reported
- Investigations completed
- Potential accident types
- Completed actions

Model 13.1 is an example of one of the many near miss incident statistics that can be reported. Model 13.2 is near miss incidents that were rectified.

EVERGREEN

As with most safety systems, the near miss incident recognition, reporting, investigation, and remedy (NEMIRR) system must be kept evergreen and not be allowed to stagnate and die out. Ongoing encouragement should be given to reporters and, most importantly, the reported hazards must be fixed or the system will lose momentum and credibility. This is not a new gimmick in safety. The system is not a safety flavor of the month. Near miss reporting and rectification has been a part of safety philosophy for a long time, so it is not something new. Ongoing support from the leadership is essential and near miss incident recall should be the opening item on every safety meeting's agenda. Eventually, reporting of near miss incidents, hazards, and safe work will become part of the safety culture. Employees and management will not think twice to identify a situation, complete the reporting form, and place it into the system.

MAIN REQUIREMENTS

There are many factors needed in order for near miss incident reporting to be effective. These include a willingness by the organization to learn from its mistakes, open and honest communication between workers and supervisors, and the removal of fear of reprisal for reporting. There should be positive motivation to report near miss incident situations and positive actions must be taken immediately to eliminate the problems reported.

This must be followed by regular feedback on reported issues and progress on rectification actions.

THE MAGMA COPPER CASE STUDY

The following case study is from the Magma Copper Company's Safety Observation Program (Gessner, 1998).

A basic, yet powerful, tool for risk management was the Safety Observation Program. This was an employee-created and driven behavior-based program in which safety observations were performed at frequent intervals during the day by all employees. The average number of observations reports was 50 a day.

The observations were categorized into four sections: (1) at-risk actions, (2) unsafe conditions, (3) safe events, and (4) near miss incidents. The observations were noted on a preprinted observation form. The feedback to the employee observed, the corrective action taken, or requested corrective action, is noted on the form.

All observations were discussed and tabulated in morning departmental meetings. Business team meetings discussed the observations at management level on a daily basis along with production goals and job planning. Any incomplete observations or requests for action were logged on a departmental Safety Commitments board and monitored until they were completed.

This system provided positive feedback for initiators of reports and participants and provided a high degree of rigor around the completion of actions. The focus of the program was on the recognition and reward of safe behaviors and the identification and elimination of unsafe conditions and acts. The reporting of these accident precursors placed the organization in a proactive, forward-driven position in risk reduction efforts, on a daily basis, on the shop floor.

The program allowed people to deal with the environment in which they worked. One of the main spinoffs of the program was that it created a dialog between management and employees. An added benefit was having managers do what they say they were going to do to show their commitment to the safety of their employees. The observation program created responsibility to rectify unsafe situations and was a method to hold responsible managers and employees accountable for the corrective actions.

THE PHOENIX WATER SERVICES DEPARTMENT CASE STUDY

The following case study is from the City of Phoenix Water Services Department's Near Miss Program (Nevitt, 2011), which is one element of their Five-Star Safety and Health Management System consisting of 73 elements.

Based on the concept that for every serious injury there are 600 near misses that occur prior to the injury, the City of Phoenix Water Services Department (PWSD) implemented a near miss program as part of its Safety Management System.

Under the Five Star approach, reporting, investigating, and addressing near misses provides opportunities to evaluate the unsafe acts and unsafe conditions that can lead to losses before they actually occur. In 2010, the PWSD reported 10 serious injuries. This correlates to more than 6,000 near misses or missed opportunities to correct the unsafe act or condition and prevent the injuries.

The components of the PWSD's Near Miss Program include three sections: reporting, investigation, and correction. The PWSD developed a reporting form that enables an employee, supervisor, or manager to report an unsafe act or unsafe condition that could lead to an injury, property damage, or loss of process. The form also provides the employee with an option to indicate the potential severity (magnitude of the loss), the likely recurrence of the near miss, and any action taken to eliminate the unsafe condition or act.

Initially, this program was met with significant resistance from employees and supervisors alike. Employees indicated that identifying hazards was the "supervisor's job" and that reporting unsafe acts was like "tattling on" their co-workers. More importantly, employees expressed concerns that they would be retaliated against by their co-workers and supervisors for reporting unsafe conditions or acts or disciplined for conducting an unsafe act. Supervisors argued that the near miss program was just going to create a lot more work for them. They argued that investigating near misses would detract from their core responsibility of supervising employees. They also worried about being disciplined for unsafe conditions and acts that were reported but not investigated or corrected in a timely manner.

To address these concerns and encourage use of the near miss program, the PWSD safety and training staff met with employee work groups and supervisors to discuss their concerns and incorporate their ideas and suggestions into the final standard. Under the standard, the employee may hand the form to his supervisor or drop it in a near miss box anonymously. Supervisors are responsible for ensuring that there is no retaliation for reporting a near miss. These concerns also are discussed in the training where it is stressed that the purpose of any investigation, whether for a near miss or a loss, is to determine the facts not to find fault. This approach is reinforced in the PWSD's Health and Safety Policy.

Near miss training was provided to all employees, supervisors, and managers prior to rolling out the near miss program. The goal of the training is to introduce the concepts of the near miss program and the benefits of reporting and investigating near misses before there is an actual loss. Additionally, this course covers how to use the form to report a near miss, how to evaluate the potential severity and likely recurrence of the unsafe act or condition, and how to investigate the near miss. At the completion of the course, employees receive a near miss reporting booklet. The Five Star Standard for the near miss program and a training brief is available online if the employee or supervisor has any questions on how to use the form or conduct the investigation after attending the class.

Once the near miss has been reported, it is the supervisor's responsibility to evaluate the severity and probability of the loss. Near misses with a medium-to-extreme potential severity or a medium-to-extreme likely recurrence are required to be investigated by the supervisor. As part of the investigation, the supervisor shall identify the root/basic causes of the unsafe act or condition, and then implement corrections to prevent recurrence. The near miss program includes a standardized investigation form that aids the supervisor in the investigation and documentation of the near miss.

Corrections beyond the authority of the supervisor are forwarded to a safety committee for review and implementation. To facilitate discussion of near misses, supervisors are responsible for tracking reported near misses and posting information on the findings of the investigation and the corrective actions taken. Additionally, the PWSD's Safety Section is copied on all near miss reports to facilitate assistance, where appropriate, and to ensure sufficient follow-up action is taken.

Before the introduction of the near miss program, near misses were talked about informally amongst employees, but not formally addressed by management until there was an actual loss. Near misses were considered just part of the job of working in the water and wastewater industry. Since the implementation of the program two years ago, only 75 near misses have been reported. While the near miss program has not resulted in the reporting of a substantial number of near misses yet, the program is being used throughout the PWSD and it is helping to change the safety culture of the department. As employees see changes based on what they have identified and reported, their confidence and their role in the entire Five Star Program increases. Surprisingly, employees are not only using the form to report actual near misses, but to report unsafe acts and unsafe conditions even before there is a near miss. Supervisors have expressed an appreciation for the near miss program in that it is encouraging all employees to be accountable for their own safety and aware of their co-workers' safety. According to the supervisors, the employees are becoming more attuned to working safely and are actively looking out for unsafe conditions and acts and often times correcting them without having to rely on the supervisor.

One near miss that was reported involved a vehicle exit from a PWSD facility. Employees exiting the facility did not have a clear, unobstructed view of oncoming traffic because of landscaping. Through an investigation of the near miss, the supervisor made and implemented a recommendation to post traffic signs and restripe the exit and roadway. These minor improvements ensure traffic exiting the facility come to a stop with a clearer view of oncoming traffic before they enter the roadway.

Another near miss involved a crew cleaning debris from a roadway outside of their facility. The winding roadway creates several blind spots for vehicles traveling on it. As a result, a vehicle came around a bend in the road and nearly struck an employee working in the road. The near miss investigation resulted in the retrofit of a vehicle with a large traffic arrow board and strobe lights that the employees can now position along the roadway to warn motorists of the crew working in the road.

As a result of success stories like these, the near miss program is becoming an integral part of the PWSD's overall Five Star Health and Safety Program. Near misses are being discussed at monthly facility safety meetings as well as department safety meetings and changes are being made based on safety concerns identified by the employees. While the PWSD has not seen a decrease in losses since the near miss program was implemented, there has been a shift in the safety culture by empowering employees and supervisors to take responsibility for identifying and correcting unsafe conditions and unsafe acts as part of their core job duties. "Safety is everybody's baby."

CONCLUSION

The reporting of safe work of near miss incidents and hazards, and the rectification thereof is a vital part of any safety program and, if used correctly, can contribute more to a safety success than many other safety system elements. Heeding accident warning signs by reacting on near miss incidents gives an organization an opportunity to lever its preventative safety efforts proactively.

14 Investigating High Potential Near Miss Incidents

NO DIFFERENCE

The investigation of an injury, loss-producing accident and a near miss incident with high probability of recurrence and high loss potential is essentially the same. The only difference is that in the case of an accident one determines what happened and in the case of a near miss incident, *what could have happened*. Therefore, the difference is the actual losses created by the accident and the potential losses that could have been caused by the near miss incident, under slightly different circumstances. The investigation processes are almost identical as high potential near miss incidents should be treated with the same rigor as accidents.

PURPOSE

The purpose of both accident and near miss incident investigation is to carry out an investigation into the undesired event, to determine what happened, and what can be done to prevent a similar event recurring. If positive preventative measures are not taken after an accident or near miss incident, the probability of recurrence is great. The investigation procedure would lead to the root causes of the event and help determine what steps should be taken to prevent a similar accident or near miss incident.

ACCIDENT/NEAR MISS INCIDENT INVESTIGATION FACTS

- Accident/near miss incident investigation is a problem-solving technique.
- Investigation requires management skills.
- Approach investigations with an open and objective mind.
- Accident and near miss incident investigation are fact-finding, not fault-finding.
- Delve deeper—beyond the apparent reasons—to the underlying causes.
- Often, more information is required to obtain all the facts.
- Investigators cannot accept all the answers given at face value.
- Get statements from witnesses.
- It is important to determine and assess the potential for loss as soon as possible.
- Look beyond the injured person.
- Consider the perception of others who often see things differently.

149

- Consider communication skills and questioning techniques.
- Above all, get as much factual information as possible to get the full picture.

POSTCONTACT VERSUS PRECONTACT

Accident investigation is a *postcontact* control activity as a loss must occur before the accident can be investigated and preventative steps taken. Although a postcontact control activity accident investigation leads to the taking of preventative measures, which, in turn, is a precontact activity.

Near miss investigation is more proactive as it takes place before a loss has occurred and is a *precontact* control activity. The flow of energy has not caused damage or injury, but the accident symptoms are the same. It is easier to investigate a near miss incident because, since there has been no injury, there is less likely to be a cover up. Because there has been no loss, there is generally less of a postaccident fear factor.

POTENTIAL LOSSES AND RISK RANKING OF PROBABILITIES

Near miss incidents with high probability of recurrence and high or medium-high severity potential should be investigated and treated with the same urgency as loss-producing accidents.

MISLEADING

As a young safety advisor, I advised a number of managers and safety professionals to investigate every single near miss incident. I was taught to believe that each near miss incident could have ended up as an injury. What I was doing was misleading my clients. If I had thought through the process, I would have realized that, due to manpower and cost constraints, it is virtually impossible, as well as totally impracticable, to investigate *all* near miss incidents. Many near miss incident programs have failed because of this incorrect approach.

POTENTIAL

In tackling the no-contact events, the keyword is potential. What did the event have the capacity to do under slightly different circumstances? The potential is what should determine which of these near miss incidents should be investigated as well as the level of the investigation.

POTENTIAL HAZARDS

Safety practitioners have often debated the expression "potential hazard" and have come to the conclusion that there is no such thing as a potential hazard. They argue that a hazard is a hazard is a hazard. They are correct. The term *potential hazard* really refers to a hazard that has the potential to cause harm. The keyword here is having potential or having the capacity to cause harm.

Loss Potential

As W. E. Tarrants (1980) puts it:

> The existence of an injury or property damage loss is no longer a necessary condition for appraising accident performance. It is now possible to identify and examine accident problems "before the fact" instead of "after the fact" in terms of their injury-producing or property-damaging consequences. This allows the safety professional to concentrate on measurement of loss potential or near misses and remove the necessity of relying on measurement techniques based on the probabilistic, fortuitous, rare-event, injurious accident (p. 319).

R. P. Boylston (1990) refers to incidents as potential problems and also quotes the luck factors:

> Failures by an organization to recognize, evaluate, and implement controls for early warnings of potential problems usually result in a system of reactive approaches. Consequently, there is little if any way to control the magnitude of the problem. Such organizations are "lucky" or "unlucky," depending on the situation. This is no way to manage an organization (p. 103).

Ranking the Potential

The most important aspect of a near miss incident is the quantification and ranking of the incident's degree of potential. The degree of potential could be:

- Severity of loss (injury, property damage, environmental harm, etc.)
- Recurrence of event (How often could it happen?)
- Number of people affected (How many are exposed and how often the task is done?)

These are the common terms of probability, severity, and frequency. A simple method of ranking the potential of a near miss is to ask the following questions:

- What is the probability of this event occurring?
- If this event occurs, how bad will the consequences be?
- If the event occurs, how often will it be repeated and how many people are exposed?

Safety Solution

It is strongly felt that the solution to safety problems lie within those near miss incidents with high potential because they are accidents that the organization has not yet experienced. The loss causation sequence has been triggered, but, due to Luck Factor 1, has ended in a warning, a close call—a near miss incident. All an organization needs to do is to identify the potential of the near miss incident. If the potential is high in terms of severity and probability, then it should be treated as if something *had* happened. Investigate the event and institute appropriate control measures. Only one link in the accident chain was missing, which determined the difference between an accident and a near miss incident.

CRYSTAL BALL

A very simple method of assessing the potential of each near miss incident is to gaze into an imaginary crystal ball and quote the magic words: "It's not *what* happened, it's what *could have* happened." This very simple, potential assessment technique will identify those near miss incidents with the greatest potential for loss, which should be investigated and treated with the same urgency as loss-producing accidents.

SUCCESSFUL ASSESSING AND ANALYZING

A simple method of assessing and analyzing the potential of near miss incidents:

- Train and encourage employees to report all near miss incidents, irrespective of their potential.
- There should be anonymity of reporting.
- Assure employees that there will be no repercussions and that the system is a "no names no pack drill" exercise (amnesty).
- Commend employees on submitting near miss incident reports.
- Issue a simplified near miss incident/accident investigation form that includes a risk matrix.
- Using a simplified risk matrix system, rank the probability of recurrence and potential severity of each near miss incident.

RISK MATRIX

The following model is a simple risk matrix that rates the probability of recurrence and potential severity from low to high (Model 14.1). The near miss incidents, or high risk acts and conditions reported that fall into the gray or intermediate area, should receive investigation. Those that fall into the black (high-high or medium-high, medium-high) areas, should be investigated as thoroughly as accidents that have resulted in a loss. The same accident investigation report should be used and the same diligence applied even though there was no loss. The potential for loss is the guiding factor in this instance.

BENEFITS OF ACCIDENT AND NEAR MISS INCIDENT INVESTIGATION

Properly conducted accident and near miss incident investigations, help identify and quantify the losses (and possible losses) incurred by undesired events. They help determine the facts of the situation and also define which accident or near

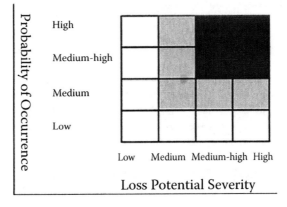

MODEL 14.1 A simple risk matrix.

miss incident type occurred. The investigation also helps determine the exact extent of the personal injury or disease, the extent of damage as well as what actually happened, or what could have happened. Only by investigating the event can the physical and occupational hygiene agencies and the agency parts be identified and classified.

Near miss incident and accident investigation also indicates who was involved at the time of the event and, most importantly, is a means to identify the immediate causes in the form of high risk acts and high risk conditions. Once these are determined, they enable us to establish the root cause or basic causes of the occurrence. Near miss incidents reveal the same information as accidents because the same factors that contribute to them are present in the near miss situation.

The immediate causes of accidents and near miss incidents are the high risk acts and high risk conditions and once determined lead to identification of the root causes in the form of *personal factors* and *job factors*.

PERSONAL FACTORS

The personal factors could be stress, lack of skill, lack of knowledge, inadequate motivation, physical or mental shortcomings, or a poor attitude toward safety. Many investigations focus on the personal factors in an effort to find fault and tend to ignore the environmental job factors.

The investigator should remember that accidents have multiple causes and the investigation should not cease when one cause has been identified. Immediate cause analysis by means of inspection, examination, and interviews should continue until all the contributing high risk situations are identified.

Once the immediate causes have been identified, a root cause analysis can be done to derive the root causes. A simple but effective method to do this is to take each immediate cause in turn and ask: "Why? Why? Why?" The answers will reveal the root causes.

As I wrote in *Changing Safety's Paradigms*" (McKinnon, 2007):

> Until accident investigation is used as an effective tool to identify the true causes, it is a waste of time. Poorly investigated accidents are missed opportunities to take positive steps to prevent recurrences. Every accident has multiple causes. Proactive safety means investigating thoroughly and identifying all the immediate and root causes of the accident. Then it involves asking why each unsafe act was committed. It further involves asking why each unsafe condition existed. This is opening a can of worms and delves into the basic causes. Remember, the employee who was injured is only a victim of the safety system failure (p. 124).

As the Columbia Accident Investigation Board (British Standards Institute, 1999) reported:

> Causal factors for accidents that result in severe injuries are multiple and complex, and relate to several levels of responsibility (p. 23).

JOB FACTORS

Job factors are the factors arising from a high risk work environment. Job factors could include items such as excessive wear and tear, inadequate safety standards, tools and equipment that are insufficient, poor or no maintenance, no standards for purchasing, lack of maintenance or poor supervision, etc.

PRINCIPLE OF MULTIPLE CAUSES

The principle of multiple causes is of paramount importance in accident and near miss incident investigation and states: "Accidents and other problems are seldom, if ever, the result of a single cause." Very rarely does an accident or near miss incidents occur as a result of a single cause. Experience has shown that multiple causes are almost always present in most accident and near miss incident situations. Accident investigation helps determine which causes contributed to the loss.

PRINCIPLE OF DEFINITION

Another important principle to remember in investigation is the principle of definition, which is: "A logical and proper decision can be made only when the basic or real problem is first defined. (Prescription without diagnosis is malpractice.)"

THE GOLDEN RULE OF ACCIDENT AND NEAR MISS INCIDENT INVESTIGATION

The golden rule of accident and near miss incident investigation is: "Accident and near miss incident investigation is fact-finding and not fault-finding."

In other words, use accidents and near miss incident investigation systems and methodology to get to the facts, the root causes of the accident, and don't use it to find fault. Finding fault will not help in identifying and removing the cause of accidents and near miss incidents.

WHO SHOULD INVESTIGATE?

The supervisor, foreman, or manager directly in charge of the area in which the event occurred should be the prime investigator. In some instances, the safety and health representative may have been appointed as the event investigator, but this should not free the immediate manager of his/her responsibilities to investigate the accident or near miss incident.

Always delegating this responsibility to the safety coordinator or safety department is not acceptable. The safety coordinator should only assist the line management in investigating the event because he/she is an expert in this field and can be used as a resource.

The extent of the loss should not determine who should carry out the investigation. Past experience has shown that once there are fatalities, multiple fatalities, or great financial loss as a result of an accident, only then does top management get involved. What should determine participation in the near miss incident or accident investigation procedure is the potential for loss of the event, or what could have happened under slightly different circumstances?

It should be remembered that the difference between a near miss incident with no loss and an accident with great loss is only a matter of luck. The undesired event or near miss incident should receive as much attention as the accident that resulted in severe injury.

INVESTIGATION COMMITTEES

In some cases, it might be beneficial to appoint a committee or subcommittee to investigate certain accidents and high potential near miss incidents. These subcommittees could use the small group activity concept or brainstorming method to identify the root causes of the event and also to propose remedial measures and actions to take to prevent recurrence.

INVESTIGATION FORM

The accident/near miss incident investigation form is the key document in the investigation process. This form could be used to investigate injury-producing accidents, high potential near misses, property damage events, fires, and environmental events.

The accident/near miss incident investigation form, which is filled out during the investigation, must prompt the investigator to get all the facts, list all the contributing factors, and provide a complete description of the accident as well as allowing for the selection of possible remedies. The form should take cognizance of unskilled investigators and, therefore, most of the information concerning accident investigation must appear on the form. This would include the listing of the immediate causes, the root causes. and also the possible remedies. These boxes can then be checked eliminating much guesswork on behalf of the investigator.

The form should contain at least the following information:

- Name of person or persons involved.
- Department where the accident/near miss incident took place.
- The part of body injured or part of body that could have been injured.
- Description of damage or effect on person or possible damage and effect.
- The losses or potential losses.
- The potential or actual accident type (contact).
- The description of what happened.
- Which high risk act or acts were committed?
- What high risk condition or conditions were present?
- What personal factors contributed to the acts?
- What job factors contributed to the immediate causes (conditions)?
- A risk evaluation of the loss severity potential, probability of occurrence, and the frequency of the exposure.
- The direct cost and total cost of the losses or potential losses.
- Possible remedial measures.
- The follow up dates.
- A cross reference to other documents concerning the accident.
- Place for signatures of responsible people and dates.
- Safety and health committee's recommendations.
- Preventative measure action plan, responsibility, and date.

The above are the basics that should appear on the investigation form. Other items, such as names of witnesses, safety program elements that could have prevented the event, etc., can be added.

The main criterion for the investigation form is: Does it give a complete and full description of:

- What happened?
- The losses?
- The immediate and basic causes of the accident?
- Positive steps to instigate the prevention of a recurrence?

NEAR MISS INCIDENT/ACCIDENT INVESTIGATION PROCEDURE

Immediately upon the occurrence of an accident or high potential near miss incident certain steps should be taken. These include:

- Immediate actions
- Gathering the facts
- Determining the causes
- Taking remedial action
- Following up

IMMEDIATE ACTIONS

Immediately upon the occurrence of an accident, the supervisor or person responsible for the area or operation should take charge of the scene and ensure that the necessary medical attention and evacuation is initiated. Medical treatment of injured persons is a priority.

Steps should be taken to ensure that no secondary accidents occur. Where possible, the accident scene should be photographed and the positions of equipment marked or noted. Witnesses in the vicinity should be identified and, where possible, the environmental conditions should be noted. That would include the state of the ventilation, lighting, ambient noise level, ground conditions, etc. The reason for noting the environmental conditions is that they may change quickly and mislead the accident investigation team.

The accident should then be reported to the people concerned and, if required, the necessary legal report must be submitted as well as the claim to Workers' Compensation or insurance. The exact outcome of the accident should be determined as it could be a disabling injury, a minor injury, a fatality, or a property damage accident.

The accident information should be made known via means of the company safety newsletter, safety Web site or similar communication means. Some organizations issue a "loss announcement" after serious injury resulting accidents and high potential near miss incidents. The information also could be posted on notice boards in the plant to inform employees of what happened.

The accident investigation is initiated as soon as possible after the accident. The sooner the investigation begins the more effective it will be. Don't leave accident investigation to two or three days after the event because the facts would have become distorted and the witnesses may not remember as clearly as they did immediately after the accident.

GATHERING THE FACTS

The accident investigation will now get the facts as to which people were involved, what equipment was involved, what were the environmental conditions concerning the lighting, noise and work situation, and also what procedures and policies were there that could have prevented the accident. These procedures could be the safety induction training, the staff selection process, the observation of critical tasks being carried out, or the written safe work procedures, among others.

Once all the facts have been obtained from the witnesses, the environment, and the positioning of equipment and machinery, the facts are then all recorded on the accident investigation form. Where necessary, witnesses' testimonies are recorded separately and cross-referenced to the form.

Getting the facts should involve the taking of photographs, taking measurements, examining items, and reviewing standards, procedures, and past experience. It is important to keep an open mind during the gathering of these facts because one should not jump to conclusions; conclusions should be derived from the facts.

DETERMINING THE CAUSES

When determining the causes of an accident, don't jump to conclusions. Don't be predictive. Rather, deduct the facts by a systematic identification of immediate causes from which the root causes can be derived.

There are two categories of immediate causes: the high risk work conditions and the high risk acts of people. These should be categorized and reproduced on the form for easy selection by checking the relevant boxes. Bearing in mind that there is always more than one cause for an accident, the high risk acts and high risk conditions must then be identified by:

- Physical inspection
- Interviewing witnesses
- Positions of the equipment and machinery

Once the high risk acts and high risk conditions are identified, ask the question: Why? By asking "why" a high risk condition existed, we will then get to the root causes of the condition. By asking "why" the person committed a high risk act, we will derive the root causes of the act. The word *causes* is used when discussing root accident causes as there are always more than one root cause identified after an accident. In many cases, there are multiple root causes that contribute to a high risk act or condition. Root cause analysis is a specialized investigation technique that should be applied to all investigations.

The root causes are categorized as personal and job factors. Only once the root causes have been identified can remedial measures be taken. Fixing the immediate cause is treating the symptom rather than the cause.

TAKING REMEDIAL ACTION

Once the facts have identified the immediate and root causes of the accident, we now need to initiate remedial steps or preventative action to prevent recurrence. An accident will not wait for these steps to be taken; therefore, urgency is of the essence.

The remedial steps may be:

- Engineering action
- Environmental action
- Training
- Reinforcing of safety rules
- Reviewing safety rules and procedures
- Setting standards
- Enforcing standards
- Disciplinary control (positive not punitive)

The engineering action could be:

- Fixing something

- Modifying the equipment
- Fitting a device
- Cleaning up an area
- Repairing the equipment
- Covering the opening
- Removing the hazard
- Replacing or substituting the materials and equipment

The environmental action could involve removing the hazard from the environment by modification or substituting materials and processes.

The other remedial steps speak for themselves, except for discipline, which should always be reviewed as the absolute last resort. Positive discipline should always precede any other form of discipline. Discipline should only come after efforts have been made to determine why the person acted the way he/she did. Perhaps, it was condoned practice? It may be the work culture? There may have been conflicting demands? Whatever the investigation uncovers, a thorough root cause analysis should be conducted before any decisions to focus on the employee's "behavior" is considered.

Discussing accident causes in *Changing Safety's Paradigms,* McKinnon (2007), I stated:

> By blaming employees for accidents, we are really over simplifying the safety problem. By making statements such as, "The majority of accidents are caused by unsafe acts of people," we are diminishing the importance of a safe work environment. We are also watering down the impact that good management has on the reduction of accidental losses. This is one of safety's sacred cows and few have the heart to challenge it. Many people say that a person must have created a high risk condition anyway. Because of this, they say, "Well, all hazards have to do with people, therefore, people are responsible for the majority of accidents." This is far, far too simplistic. The actual causes of accidents within an organization can only be determined by a careful study of all the facts, and data, and by compiling statistics applicable, and site specific, to that organization and by uncovering the root causes (p. 57).

Numerous safety programs fail because discipline is applied immediately after an accident has taken place. Discipline is also applied for near miss incidents that occur. This leads to people clamming up and not reporting both near miss incidents and accidents for fear of disciplinary action.

Whatever remedial action or steps are implemented to prevent the recurrence of an accident the main criteria is: "Will this prevent a recurrence of this accident?" If the answer is affirmative, then the action has been effective.

In some instances, it may be advisable to table the facts at the safety committee meeting and ask the committee's opinion of what should be done to prevent the accidents. The immediate supervisor in the area as well as his team is also a good resource of remedial measures. They should be included in the decision-making process.

FOLLOWING UP

Once the remedial steps have been identified, they must be implemented. The only way to ensure that they are implemented is to follow up by:

- An inspection of the work area
- A review of the necessary documentation concerning training, etc.
- A check on the effectiveness of the training
- Monitoring the standard implemented
- Observing the new procedure

Follow up must involve action. Follow-up action must ensure that what steps were recommended, in fact, has been implemented. The efficiency of the remedial steps also should be inspected to ensure that the remedy is effective. Prescription of a remedy without correct diagnosis is safety malpractice.

LOST OPPORTUNITIES

Any form of monetary award or bonus that is linked to the degree of injury severity should be discontinued as this would lead to a cover up of damage and injury-causing accidents and other loss events. Thus, they will remain hidden and will not be investigated and the organization will not be able to learn from its mistakes. Discipline should never be implemented after an accident because it is absolutely the last resort to prevent accidents from recurring. Disciplinary measures will drive future injury accidents underground and they will not be reported or investigated. Opportunities to fix weaknesses in the safety and health system will be lost.

CONCLUSION

All accidents that cause injury, damage, or business loss must be investigated to determine what happened so that preventative measures can be initiated to prevent a recurrence. All near miss incidents that have high potential to cause loss should receive the same attention as accidents and also should be thoroughly investigated in the same way and with the same gusto as accidents.

Accident investigation should be fact-finding and not fault-finding and, if used effectively, can contribute greatly to the reduction of accidents and near miss incidents occurring in a workplace. Once the near miss incidents are identified and their causes corrected the probability of accidents occurring is reduced.

15 Summary

SAFETY IN THE SHADOWS

Near miss incidents have often been referred to as "Safety in the Shadows," because this is where the heart of the accident problem lies. Near miss incidents offer management an opportunity to rectify a system breakdown before an accident happens. Because there are no losses as a result of an undesired event does not necessarily mean that the event is insignificant. Risk assessment of near miss incidents will determine which ones warrant a full investigation. This will help to track and eliminate the source of the problem at the root.

Just because an injury is minor does not mean that the event that caused the injury was minor. The event should be investigated and the potential and probability of recurrence evaluated. The next similar event may have far worse consequences due to luck factors. (Remember, under slightly different circumstances ...)

SUMMARY OF THE BOOK

IDENTICAL CAUSES

The causes of near miss incidents and accidents are identical. The only significant difference is the outcome or consequence. An accident causes a loss and a near miss incident does not result in a loss. The missing link in the accident chain is the exchange of energy resulting in a near miss incident and not an accident. History of thousands of undesired events has shown that the outcome of the event cannot be predicted and that, under slightly different circumstances, the consequences could have been worse if it were not for luck factors.

INSUFFICIENT ENERGY

Some near miss incidents come close to causing some form of loss as there is an actual exchange of energy. The exchange is insufficient to cause loss or injury, but the fact that there was an exchange of energy is reason enough to heed the warning. Remember, it's not what happened ... but what could have happened.

As S. L. Smith (1994) said, "If enough near misses occur, the question is not, will an accident happen, but when will it happen" (p. 33).

LOSSES

Each accident results in some form of loss, and all losses cost money. Time may be lost, forms need to be filled out, and the business is interrupted to a degree. Many of the costs of an accident are hidden and, therefore, go unnoticed. Direct costs or insured costs are normally the only costs associated with an accident. They are the least of the three costs. The second level of cost is the indirect costs and a third, deeply hidden layer of costs is the totally hidden costs, which are seldom identified or tallied.

ACCIDENT RATIO

International research by many safety practitioners has identified a definite ratio between injury-related accidents and near miss incidents. Each serious injury-related accident is an indication that there have been some minor injury-related accidents, more property damage accidents, and plenty of near miss incidents. Below the waterline are numerous high risk situations that have gone unnoticed and untreated.

One cannot focus on the tip of the accident ratio iceberg alone, as the tip is the result of the base, which, in the accident ratio, is the high risk conditions and behavior that lead to the losses. Focusing on the tip (the serious injuries) is treating the symptoms of the problem and not the cause. Once action is taken on the base of the iceberg, the injuries that make up the tip are reduced.

Management's biggest opportunity for injury reduction is the many near miss incidents that indicate, in advance, weaknesses in the safety management system.

MULTIPLE CAUSES

The principle of multiple causes indicates that accidents and near miss incidents are usually the result of multiple causes. Investigations should not cease until all the contributing causes have been identified. Once all the obvious causes are found, a root cause analysis should be conducted to delve into the root causes of the problem. Only by identifying and rectifying the root causes will the problem be solved.

Immediate and Root Accident Causes

High risk acts and conditions are the immediate causes of accidents. They are the obvious causes, or the causes that lead to the contact with a source or sources of energy. Immediate causes are the result of often deep-lying root causes.

A structured system of investigation identifies the high risk acts of the employees and high risk work environments, but also asks what the root causes were for these high risk situations. The high risk acts and conditions are only symptoms of a failure in the management system. There is always a reason why a person behaves the way he/she does. If a person behaves in a high risk manner, it may be because his/her local supervisor is not supervising the staff correctly. Therefore, the behavior is tolerated, and, as such, condoned. This would be a root cause. Treating the immediate causes is treating the symptom of the safety problem, not the root cause. Root cause analysis will identify the reasons behind the high risk behavior and conditions. These are what need to be remedied to eliminate possible future accidents.

REPORTING NEAR MISS INCIDENTS

Reporting all undesired events, such as near miss incidents, is perhaps the most important aspect of any safety management system. A no-blame system should be introduced to encourage reporting without consequence. Most near miss incident programs fail as a result of disciplinary steps being taken once an event has been reported. The more warnings that are turned in, the more the opportunity to investigate, identify, and rectify the root causes of accidents before they happen.

Employees should be encouraged to report any safety issues, such as high risk work conditions and behavior as well as safe work. Having employees report safety deviations is almost like having a workforce of safety inspectors.

Training sessions on the basics of near miss incident philosophy and benefits, as well as how and what to report, should accompany the introduction of the system.

A suitable form and other reporting methods should be provided before introducing a near miss incident reporting system and a management standard should be published indicating responsibilities and accountabilities for the various aspects of the system.

Acknowledgment of receipt of the report should be forthcoming and the follow-up actions taken or where they are delegated should appear on some form of master tracking sheet that is publicized on a regular basis.

RISK ASSESSMENT

Not all near miss incidents have high potential to cause injury and loss, yet some do. The only way to prioritize the reported occurrences is to risk-rank them by means of a risk assessment. The best tool for this is the risk matrix. Remember, it's not what happened, it's what could have happened. The risk matrix is a crystal ball to predict the future or possible outcome of an event. Use it to forecast the probability of the next loss. Near miss incidents that fall into the high–high areas on the risk matrix should receive priority for investigation and rectification.

NEAR MISS INCIDENT INVESTIGATION

High loss potential events, even though they did not result in loss, should be investigated as rigorously as serious injury-producing accidents, i.e., if the assessment of the risk shows in the high–high areas on the matrix. Risks that rank in the gray area, or medium level, also should be subject to an investigation. All injuries and loss-producing events (in excess of $1,000) should be investigated irrespectively. Investigation should be a fact-finding mission to uncover the root causes of the event and not a fault-finding exercise.

SAFETY MANAGEMENT SYSTEM

The near miss incident recognition, reporting, ranking, and remedy (NEMIRR) system is not a behavior modification program. Near miss systems can only play a part in the total safety management system. It can only have a role once the safety system is in place and is monitored constantly by ongoing audits, checks, and balances.

Sometimes safety practitioners, without an in-depth knowledge of the workings of the safety management function, tend to grab any "new approach" to safety. These are normally in the form of some gimmick or other "flavor of the month" approach to safety. Reporting near miss incidents is not a flavor of the month or a safety silver bullet. Getting back to basics and recognizing near miss incidents as important as actual accidents will lead to a reduction of losses. Remember, no injury does not mean that there was no accident.

AUDITING

Most safety management systems count the serious injuries as a measure of "safety." This measurement method, while still accepted, is a measure of failure. Auditing the activities that make up the control measures is the best way to measure safety. This result is a more positive measurement of management work being done to control loss. The efficiency and effectiveness of the safety management system, including the near miss incident reporting system, can only be measured by quantifying it against established standards by an audit.

REDUCING ACCIDENT PROBABILITY

Workplaces will seldom be injury free. No matter what degree of safety control and effort is put in, eliminating all accidents may not be possible. Safety management systems, safety campaigns, and safety programs cannot eliminate injuries if they do not reduce existing risks. Injuries, or the end resultant losses, are determined by luck factors. What proactive safety efforts do is reduce the probability and the frequency of the event occurring. An organization can make the efforts to reduce the chances of it happening by identifying and remedying the warning signs—near miss incidents.

Friendly Warnings

Near miss incidents are warnings that the management system is flawed. Not all near miss incidents, however, need immediate and in-depth actions as some near miss incidents have less potential than others do. Assessing the potential of a near miss incident gives a clear indication so that investigations can be prioritized. Using a simple risk matrix, each near miss incident's potential can be ranked and consequent investigations and follow-up actions prioritized.

Near miss incidents should be treated as friendly warnings. Friendly warnings that have high potential offer a clear indication of what *could* happen under slightly different circumstances, which are normally beyond our control. Management should take heed of these high potential, near miss incidents and institute controls before a loss-producing accident occurs. Should a contact and accidental exchange of energy take place, the consequence of that contact is fortuitous.

GUIDELINES

- In order for any organization to improve safety by reducing losses, it is necessary for the organization to openly share and learn from its mistakes in the form of high potential near miss incidents.
- Sharing situations that "almost" or "could have" resulted in injury or loss creates opportunities to discuss what happened and to learn from these events.
- It is very important that there be an open and trusting environment (amnesty) where all near miss incidents can be reported.
- Reporting near miss incident situations should be acknowledged, supported, encouraged, and praised.
- The system should evaluate near miss incidents by severity and potential, establish priorities, and lead to positive corrective action. The risk matrix is the ideal tool for this.
- If reporting near miss incidents results in disciplinary action or nothing is done to correct high severity and probability of potential situations, reporting will die and all related learning will stop.
- Since safety responsibility occurs at all levels, everyone must participate and share their mistakes.

CONCLUSION

Near miss incidents occur far more often than loss-producing accidents. They are warnings and should be heeded. According to Frank E. Bird Jr. (1996), for every serious injury there are 600 other occasions, which could have led to the injury. Safety can take a quantum leap by reporting, identifying, risk ranking, and rectifying those near miss incidents with high loss potential.

High potential near miss incidents indicate a failure of the safety management system and are warnings that under different circumstances a loss could be caused by a similar failure. Reducing the number of near miss incidents helps reduce the numerous high risk acts and conditions lying beneath the surface. Remedying near miss causes will lead to a reduction in injuries and help develop a positive safety culture.

References

American National Standards Institute, Inc. (ANSI). 2005. Standard ANSI/AIHA Z10–2005, *Occupational Health and Safety Management Systems.* Washington, D.C., p. 17.

American National Standards Institute, Inc. (ANSI) Guidelines. 2003. ANSI/ASSE Z590.2-2003, *Criteria for Establishing the Scope and Functions of the Professional Safety.* Washington, D.C.

American Society of Safety Engineers. Online at: www.asse.org

Aviation Safety Reporting System. Online at: http://asrs.arc.nasa.gov/overview/summary.html

Bagian, J. P. 2004. The opportunity of precursors. In *Accident precursors analysis and management: Reducing technological risk through diligence,* ed. J. Phimister, V. Bier, and H. Kunreuther, 39. Washington, D.C.: The National Academies Press.

Bird, F. E. Jr. and G. L. Germain 1992. *Practical loss control leadership*, 2nd ed. Loganville, GA: International Loss Control Institute, p. 52.

Bird, F. E. Jr. and G. L. Germain. 1996. *Practical loss control leadership*, 3rd ed. Loganville, GA: Det Norske Veritas, p. 21.

Boylston, R. P. 1990. *Managing safety and health programs.* New York: Van Nostrand, p. 103. (Reprinted with permission of John Wiley & Sons, New York.)

Bridges, W. G. 2000. Get near misses reported, process industry incidents: Investigation protocols, case histories, lessons learned. Paper presented at the *Center for Chemical Process Safety International Conference and Workshop.* American Institute of Chemical Engineers, New York, October 2–6, p. 379.

British Safety Council. (In 1974/75, the *Tye-Pearson* theory was conducted on behalf of the British Safety Council and was based on a study of almost 1 million accidents in Britain.)

British Standards Institute. 1999. OHSAS 18001 (Incorporating amendment 1).

Columbia Accident Investigation Board (CAIB) Report. Vol. 1. Washington, D.C.: NASA, August 2003. Online at: www.nasa.gov/columbia/home/CAIB-Vol1.html (July 2006).

Confidential Incident Reporting and Analysis System. Online at: http://www.ciras.org.uk/whatisciras.html

Friend, M. A. 1997. Examine your safety philosophy. *Professional Safety* February: 34.

Gessner, C. 1998. *The Magma Copper case study.* (Former Magma/ BHP Director of Safety and Loss Control; currently the AURA insurance program administrator, risk manager, assistant facilities manager, National Optical Astronomy Observatory, Tucson, AZ.) ().. (1998).

Health and Safety Executive (U.K.). 1976. *Success and failure in accident prevention*, July.

Health and Safety Executive (U.K.). 2006. (Contains public sector information published by the Health and Safety Executive and licensed under the Open Government License v1.0'.) Online at: http://www.hse.gov.uk/statistics/causinj/index.htm.

Heinrich, H. W. 1931. *Industrial accident prevention: a scientific approach.* New York: McGraw-Hill. (Out of print).

Heinrich H. W. 1950. *Industrial accident prevention: A scientific approach*, 3rd. ed. New York: McGraw-Hill Book Company, p. 10. Online at: http://lachlan.bluehaze.com.au/books/nsc_safety_facts_1941/index.html

Heinrich, H. W. (assisted by E. R. Granniss). 1959. Industrial accident prevention: A scientific approach, 4th ed. New York: McGraw-Hill, p. 24–36.

Heinrich, H. W., D. Petersen, and N. Roos. 1969. *Industrial accident prevention,* 5th ed. New York: McGraw-Hill.

Hudson, L. A. 1993. *Insights into management.* Safety Management Society.

Jones, S., C. Kirchsteiger, and W. Bjerke. 1999. The importance of near miss reporting to further improve safety performance. *Journal of Loss Prevention in the Process Industries* (January special issue on "International Trends in Major Accidents and Activities by the European Commission towards Accident Prevention") 12 (1), Elsevier Science.

Krause, T. R. 1997. *The behavior-based safety process,* 2nd ed. New York: Van Nostrand Reinhold, p. 293.

London Underground, The. 1987. A memorandum to the Operating Management meeting.

McKinnon, R. C. 1995. Safety and health at work: An introduction. (Unpublished work)

McKinnon, R. C. 2000. *The cause, effect, and control of accidental loss, with accident investigation kit* (CECAL). Boca Raton, FL: CRC Press.

McKinnon, R. C. 2007. *Changing safety's paradigms.* Government Institutes, pp. 12, 57, 124.

National Fire Fighter Near-Miss Reporting System. Online at: http://www.firefighternearmiss.com/index.php/home

National Institute for Safety and Health (NIOSH). 2011. Online at: http://www.cdc.gov/niosh/; http://www.cdc.gov/search.do?q=management+responsibility&spell=1&ie=utf8&subset=niosh.

National Occupational Safety Association (NOSA). 1990. *Effective accident/incident investigation,* Vol. HB4.12.50E. Pretoria, SA:NOSA.

National Occupational Safety Association (NOSA).1990. *We shall call it the Hawthorne Effect.* Safety Management Training Course (SAMTRAC). Pretoria, SA: NOSA.

National Occupational Safety Association (NOSA). 1991. *MBO 5-star safety management system introduction.* Pretoria, SA: NOSA, p. 7.

National Occupational Safety Association (NOSA). 1992. *Safety management.* Pretoria, SA: NOSA (September).

National Safety Council. 2010. *Summary from Injury Facts,* 2010 ed. Itasca, IL: NSC. Online at: http://www.nsc.org/search/results.aspx?k=injury facts

Nevitt, L. 2011. *The Phoenix Water Services Department Case Study.* Phoenix, AZ. Phoenix Water Services Department.

Occupational Safety and Health Administration (OSHA). 2006. Online at: http://www.osha.gov/pls/imis/citedstandard.sic?p_esize=&p_state=FEFederal&p_sic=all; http://www.osha.gov/pls/imis/citedstandard.sic?p_sic=B&p_esize=&p_state=FEFederal

Patient Safety Reporting System. Online at: http://www.psrs.arc.nasa.gov/programoverview/index.html

Peters, T. and R. H. Waterman, Jr. 1982. *In search of excellence.* New York: Harper & Row.

Petersen, D. 1997. Why safety is a 'people problem.' *Occupational Hazards* 39–40, January.

Phimister, J. R., V. M. Bier, and H. C. Kunreuther. 2004. *Accident precursor analysis and management: Reducing technological risk through diligence.* Appendix D. Washington, D.C.: National Academies Press, p. 198.

Smit, E. and N. I. Morgan, et al. 1996. *Contemporary issues in strategic management,* 1st ed. Pretoria, SA: Kagiso Tertiary.

Smith, S. L. 1994. Near-misses: Safety in the shadows. *Occupational Hazards*: 33–36.

Tarrants, W. E. 1980. *The measurement of safety performance.* New York: Garland STPM Press, p. 319.

U.S. Bureau of Labor Statistics. Online at: http://www.bls.gov/news.release/osh.nr0.htm

You Magazine. 1994. Born out of tragedy. Parklands, SA: Gallo Images (Pty) Ltd.

Wald, M. L. 2007. Fatal airplane crashes drop 65%. *New York Times*, October 1.

Wikipedia, The free encyclopedia. 2011. Online at: http://en.wikipedia.org/wiki/Management; http://en.wikipedia.org/wiki/Near_miss

Williamsen, M. 2009. Direct Delivery Leadership Council (DDLC) Conference, Las Vegas. http://ezinearticles.com/?Near-Miss-Incident-Reporting&id=3493824

Index